The Imagination of Plants

SUNY Series on Religion and the Environment

Harold Coward, editor

The Imagination of Plants

A Book of Botanical Mythology

MATTHEW HALL

Published by State University of New York Press, Albany

For information, contact State University of New York Press, Albany, NY
www.sunypress.edu

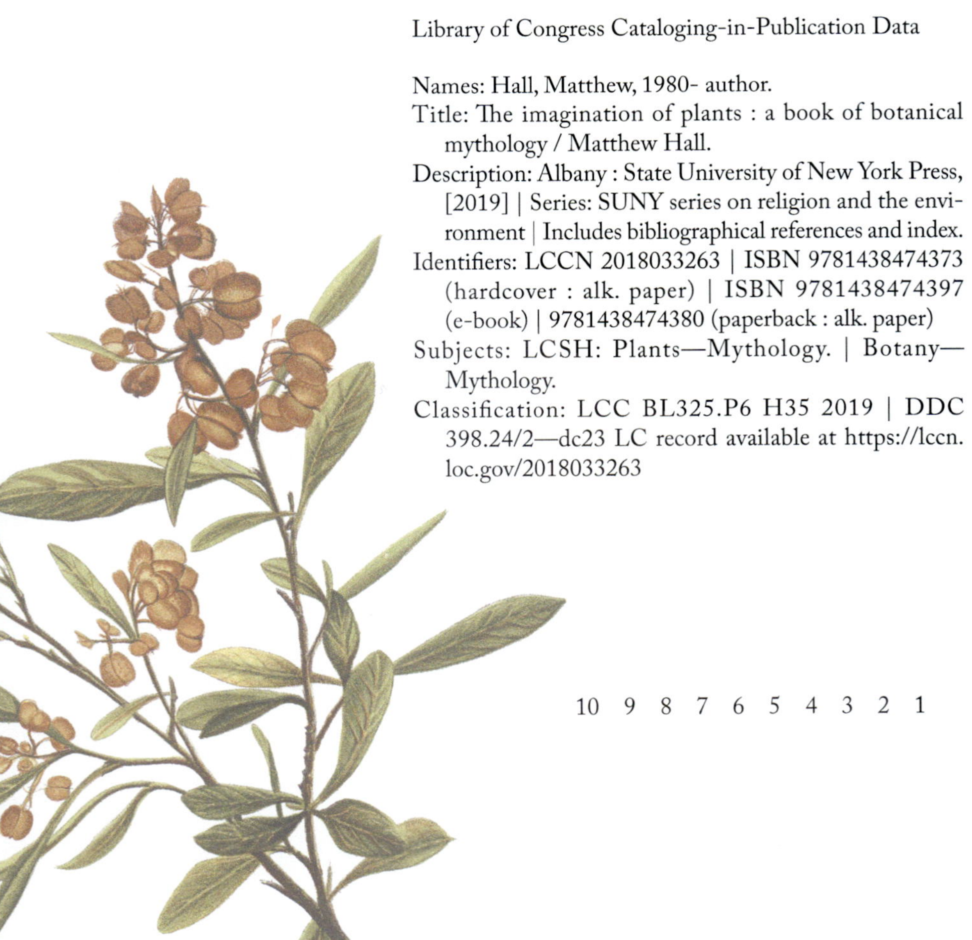

Library of Congress Cataloging-in-Publication Data

Names: Hall, Matthew, 1980- author.
Title: The imagination of plants : a book of botanical
    mythology / Matthew Hall.
Description: Albany : State University of New York Press,
    [2019] | Series: SUNY series on religion and the envi-
    ronment | Includes bibliographical references and index.
Identifiers: LCCN 2018033263 | ISBN 9781438474373
    (hardcover : alk. paper) | ISBN 9781438474397
    (e-book) | 9781438474380 (paperback : alk. paper)
Subjects: LCSH: Plants—Mythology. | Botany—
    Mythology.
Classification: LCC BL325.P6 H35 2019 | DDC
    398.24/2—dc23 LC record available at https://lccn.
    loc.gov/2018033263

For Magnolia, Iris, and Hebe

CONTENTS

ILLUSTRATIONS

PAGE I ❦ The myrtle (*Myrtus communis*), the plant into which Polydorus was transformed. Ulisse Aldrovandi, 1550–1605. Biblioteca Universitaria di Bologna, Ms. 124, *Tavole di piante*, vol. II, c. 176. © Alma Mater Studiorum Universita di Bologna— Biblioteca Universitaria di Bologna.

PAGES II–III ❦ The grey mangrove (*Avicennia marina*). From *Luigi Balugani's drawings of African plants, From the collection made by James Bruce of Kinnaird on his travels to discover the source of the Nile 1767-1773*, pl. 76 (1991), by Paul Hulton. Yale Center for British Art, Paul Mellon Collection.

PAGE IV ❦ The fruits and leaves of *Dodonaea viscosa*, known as *akeake* in Te Reo Māori. From *Indigenous Flowers of the Hawaiian Islands*, t. 39 (1885), by Francis Sinclair. Artist I. Sinclair. Image from the Biodiversity Heritage Library. Digitized by Missouri Botanical Garden www.biodiversitylibrary.org

PAGE V ❦ *Harakeke*, the New Zealand flax (*Phormium tenax*). From *Les Liliacées*, vol. 8: t. 448 (1805-1816), by P.J. Redouté. Image from the Biodiversity Heritage Library. Digitized by Missouri Botanical Garden, Peter H. Raven Library. www.biodiversitylibrary.org

PAGE VI ❦ The grey poplar (*Populus x canescens*). From *The North American sylva*, vol. 2: t. 100 (1817-1819), by F.A. Michaux. Artist P. Bessa. Photograph by Matthew Hall.

PAGE VIII ❦ The rowan (*Sorbus aucuparia*). From *Flore des environs de St. Petersbourg et de Moscou* t. 2 (1811), by J. Liboschitz and K.B. Trinius. Image from the Biodiversity Heritage Library. Digitized by Missouri Botanical Garden. www.biodiversitylibrary.org

PAGE XVII ❦ The black birch (*Betula nigra*) [as *Betula rubra*]. From *The botanical cabinet*, vol. 13: t. 1248 (1827), by Conrad Loddiges. Artist G. Cooke. Image from the Biodiversity Heritage Library. Digitized by Missouri Botanical Garden, Peter H. Raven Library. www.biodiversitylibrary.org.

PAGE XX ❦ The rowan (*Sorbus aucuparia*). From *Flora Danica*, fasicle 18, t. 1034 (1761-1883), by G.C. Oeder et al. Courtesy of The Royal Library, National Library of Denmark.

PAGE XXXII ❦ The pomegranate (*Punica granatum*). From *Curtis's Botanical Magazine* vol. 43: t. 1832a (1816). Image from the Biodiversity Heritage Library. Digitized by Missouri Botanical Garden. www.biodiversitylibrary.org

PAGE 1 ❦ Detail of the sweet potato vine (*Ipomoea batatas*), known as kumara in Te Reo Māori. From Indigenous Flowers of the Hawaiian Islands, t. 15 (1885), by Francis Sinclair. Artist I. Sinclair. Image from the Biodiversity Heritage Library. Digitized by Missouri Botanical Garden www.biodiversitylibrary.org

PAGE 14 ❦ Adam and Eve. Albrecht Dürer, 1504. Courtesy of The Metropolitan Museum of Art.

PAGE 17 ❦ Viratapurusha (Puruṣa) shown with multiple heads, standing on a pedestal with a long garland of lotus flowers on his shoulders. From *Album of 136 illustrations of the History of the world or Creation of the universe*. Karikal (Tanjore) and for some pages Masulipatam (Andhra), between 1727 and 1758. Source Bibliothèque nationale de France.

PAGE 19 ❦ The grapevine was born from the death of the sacred ox. Maria Sibylla Merian, 1701–05. Image from the Biodiversity Heritage Library. Digitized by the Smithsonian Museum. www.biodiversitylibrary.org.

PAGE 23 ❦ The common Juniper (*Juniperus communis*—here as *Juniperus minor*) is one of four coniferous species in Finland. According to the Kalevala it was sown in stony places by Pellervoinen. From *De Historia Stirpium Commentarii Insignes*, t. 78 (1542), by Leonhart Fuchs. Image(s) courtesy History of Science Collections, University of Oklahoma Libraries; copyright the Board of Regents of the University of Oklahoma.

PAGE 24 ❦ The middle column of page 9 shows the Maize God (left) speaking with the creator god Itzamnaaj (right). From the *Codex Dresden*, also known as *Codex Dresdensis*, a preconquest pictorial manuscript; ritualcalendrical. Unknown, 1500 (circa). SLUB Dresden, Mscr.Dresd.M.32, digital.slubdresden.de/id272362328.

PAGE 28 ❦ *Cycas angulata*, the cycad brought to Yanyuwa country by the Tiger Shark. From *Botanical Drawings from Australia* plate 159 (1801), by Ferdinand L. Bauer. Copyright Natural History Museum, London.

PAGE 29 ❦ *Eleocharis dulcis* (as *Scirpus tuberosus*). *From Plants of the coast of Coromandel*, vol. 3: t. 231 (1819), by William Roxburgh. Image from the Biodiversity Heritage Library. Digitized by Missouri Botanical Garden. www.biodiversitylibrary.org.

PAGE 31 ❀ The New Zealand kauri, *Agathis australis* (as *Dammara australis*). The largest specimen of the kauri tree in New Zealand is known as Tāne Mahuta, lord of the forest, from the Maori god Tāne who brought the trees to earth. From *Description of the genus Pinus and some other remarkable plants*, 2nd ed., vol. 2: t. 6 (1890), by A. B. Lambert. Image from the Biodiversity Heritage Library. Digitized by Missouri Botanical Garden. www.biodiversitylibrary.org.

PAGE 32 ❀ Mishima Pass in Kai Province. Travelers embrace the trunk of the ancient *Cryptomeria* tree (detail). Katsushika Hokusai, 1830–32. Courtesy of The Metropolitan Museum of Art.

PAGE 34 ❀ The common lime (*Tilia x europaea*). From *De Historia Stirpium Commentarii Insigne* ("Notable commentaries on the history of plants"), t. 862 (1542), by Leonhart Fuchs. Image(s) courtesy History of Science Collections, University of Oklahoma Libraries; copyright the Board of Regents of the University of Oklahoma.

PAGE 35 ❀ A South American species of passionflower (*Passiflora laurifolia*). The passionflower is symptomatic of the Christian approach to sacred plants. From *Metamorphosis insectorum surinamensium* (c.1701 – 05), by Maria Sibylla Merian. Image from the Biodiversity Heritage Library. Digitized by Smithsonian Museum. www.biodiversitylibrary.org.

PAGE 45 ❀ A South American species of passionflower (*Passiflora laurifolia*). The passionflower is symptomatic of the Christian approach to sacred plants. From *Metamorphosis insectorum surinamensium* (c.1701 – 05), by Maria Sibylla Merian. Image from the Biodiversity Heritage Library. Digitized by Smithsonian Museum. www.bio diversitylibrary.org.

PAGE 47 ❀ Leaves of *Ficus religiosa* (Peepul, papal, pipul, or Bodhi tree), unknown Indian artist commissioned by William Roxburgh, 1751–1815. Copyright Kew Gardens.

PAGE 49 ❀ Yggdrasil, with the animals that live in its branches, including Níðhöggr who bites the tree from below. From the seventeenthcentury manuscript AM 738 4to. Courtesy of the Árni Magnússon Institute, Iceland.

PAGE 54 ❀ *Ceiba pentandra*, the *yaxche* tree (as *Eriodendron anfractuosum*). From *Curtis's Botanical Magazine*, vol. 61 [ser. 2, vol. 8]: t. 3360 (1834). Artist M. Young. Image from the Biodiversity Heritage Library. Digitized by Missouri Botanical Garden. www.biodiversitylibrary.org.

PAGE 57 ❀ A priest offers springs of tulsi (*Ocimum tenuiflorum*) to Vishnu. Unknown, 1820 (circa). © Trustees of the British Museum

PAGE 98 ❧ *Narcissus poeticus*—native to Greece and named by Linnaeus in honor of its association with the myth of Narcissus. From *Les Liliacées*, vol. 3: t. 160 (1805–1816), by P. J. Redouté. Image from the Biodiversity Heritage Library. Digitized by Missouri Botanical Garden. www.biodiversitylibrary.org.

PAGE 102 ❧ *Oryza sativa*. From *Hortus Romanus juxta Systema Tournefortianum*, vol. 8: t. 64 (1783–1816), by Giorgio Bonelli. Image from the Biodiversity Heritage Library. Digitized by Missouri Botanical Garden. www.biodiversitylibrary.org.

PAGE 106 ❧ *Nelumbo nucifera*, the sacred lotus (as *Nelumbo speciosa*). From *Curtis's Botanical Magazine*, vol. 68 [ser. 2, vol. 15]: t. 3917 (1842). Artist, W. H. Fitch. Image from the Biodiversity Heritage Library. Digitized by Missouri Botanical Garden. www.biodiversitylibrary.org.

PAGE 108 ❧ The mandrake (*Mandragora officinarum*). From *De Historia Stirpium Commentarii Insigne*, t. 530 (1542), by Leonhart Fuchs. Image(s) courtesy History of Science Collections, University of Oklahoma Libraries; copyright the Board of Regents of the University of Oklahoma.

PAGE 109 ❧ *Pouteria lucuma*, the *lucma* tree (as *Achras lucuma*). From *Flora Peruviana, et Chilensis*, Plates 153–325, vol. 2: p. 17, t. 239 (1798–1802), by H. Ruiz Lopez and J. Pavon. Image from the Biodiversity Heritage Library. Digitized by Missouri Botanical Garden. www.biodiversitylibrary.org.

PAGE 121 ❧ The Barnacle Tree. From *The Herball* (1597), by John Gerard. Image(s) courtesy History of Science Collections, University of Oklahoma Libraries; copyright the Board of Regents of the University of Oklahoma.

PAGE 123 ❧ A Scythian lamb, a plant animal found in Russia (a fourlegged object composed of sticks and fur). From "Museum Britannicum," Plate 15, by Jan and Andreas van Rymsdyk, 1772. © Trustees of the British Museum.

PAGE 125 ❧ The Waqwaq tree. From a fifteenthcentury Arabic manuscript known as *Kitab al-bulhan*. Ms. Bodl. Or. 133, fol. 4lb. Courtesy of The Bodleian Libraries, The University of Oxford.

PAGE 127 ❧ Moonwort (*Botrychium lunaria*). From *De Historia Stirpium Commentarii Insignes* t. 482 (1542), by Leonhart Fuchs. Image(s) courtesy History of Science Collections, University of Oklahoma Libraries; copyright the Board of Regents of the University of Oklahoma.

PAGE 166 ❧ The myrtle (*Myrtus communis*), the plant into which Polydorus was transformed. Ulisse Aldrovandi, 1550–1605. Biblioteca Universitaria di Bologna, Ms. 124, *Tavole di piante*, vol. II, c. 176. © Alma Mater Studiorum Universita di Bologna—Biblioteca Universitaria di Bologna.

PAGE 170 ❧ Frosty Birch Trees. Akseli GallenKallela, 1894. GallenKallela was a Finnish painter best known for his depictions of scenes from the *Kalevala*. Courtesy of The Athenaeum. https://www.the-athenaeum.org/art/detail.php?ID=57558.

PAGE 178 ❧ Iskander (Alexander) at the Talking Tree. Unknown artist, 1330–36 (circa). Freer Gallery of Art and Arthur M. Sackler Gallery, Smithsonian Institution, Washington, DC: Purchase—Charles Lang Freer Endowment, F1935.23.

PAGE 182 ❧ Flowers of *Bombax ceiba*, the *salmali* tree (as *Bombax heptaphyllum*). From *Plants of the Coast of Coromandel*, vol. 3: t. 247 (1819), by William Roxburgh. Image from the Biodiversity Heritage Library. Digitized by Missouri Botanical Garden. www.bio diversitylibrary.org.

PAGE 186 ❧ Nymphs in the garden of the Hesperides picking the golden apples (detail). Cherubino Alberti (engraver), Polidoro da Caravaggio (artist), 1570-1615. Courtesy of Te Papa.

PAGE 188 ❧ The European oak (*Quercus robur*) [as *Quercus fastigiata*]. From *Traité des arbres forestiers*, t. 16, by J.H. Jaume Saint-Hilaire (1824). Image from the Biodiversity Heritage Library. Digitized by Missouri Botanical Garden, Peter H. Raven Library. www.biodiversitylibrary.org

PAGE 189 ❧ Reproductive parts (1-3. Stamens, 4-5. Cone scales, 6. Seed with wing) of the Cedar of Lebanon (*Cedrus libani*). From *Traité des arbres et arbustes*, Nouvelle édition vol. 5: t. 78 (1812), by H.L. Duhamel du Monceau. Image from the Biodiversity Heritage Library. Digitized by Real Jardín Botánico de Madrid. www.biodiversitylibrary.org.

PAGE 201 ❧ Compendium of four images from the Jaina *Uttaradhyayanasutra* showing, clockwise from top left: (top left) The six *lesyas*, external appearances cast upon the soul by the individual's Karma, represented by a roseapple tree (*jambu*) from which six men are trying to get the fruit. These men are shown by the colors of their respective *lesyas*. The black has an axe and endeavors to cut the tree down at its base. The blue is cutting the branches. The grey is cutting off the twigs with fruit. The red climbs the tree to pluck the fruit. The yellow pulls off what he can reach as he walks around. The white collects only what has fallen to the ground. (top right) Parable of the ram (*elaijjam*) and the story of the king and forbidden mangoes. (bottom left) Mahavira classifying living and nonliving things. (bottom right) The nature of karma.

All images © Victoria and Albert Museum, London.

PAGE 202 ❦ Ulysses Conducted by Calypso to the forest, cuts the trees to build his ship. Philip Dawe, 1776. © Trustees of the British Museum.

PAGE 204 ❦ Corn (*Zea mays*). From *Flore des serres et des jardin de l'Europe*, vol. 19 (1845) by L. van Houtte. Image from the Biodiversity Heritage Library. Digitized by Missouri Botanical Garden. www.biodiversitylibrary.org.

PAGE 208 ❦ The Broken Pine, Akseli GallenKallela, 1906. Image from The Athenaeum. http://www.the-athenaeum.org.

PAGE 214 ❦ *Metrosideros robusta*, the northern rātā, an endemic tree species of Aotearoa New Zealand. From *The botany of the Antarctic voyage of H.M. ships Erebus and Terror in the Years 1839–1843, under the command of Captain Sir James Clark Ross*, vol. 2(1): t. 17 (1853), by J. D. Hooker. Artist W. H. Fitch. Image from the Biodiversity Heritage Library. Digitized by MBLWHO Library. www.biodiversitylibrary.org.

PAGE 216 ❦ Krishna Counsels the Pandava Leaders, page from a *Mahābhārata* series (detail). Unknown, 1830–1850. Courtesy of the Brooklyn Museum – CC BY 3.0.

PAGE 220 ❦ *Cedrus libani*, the cedar of Lebanon, (as *Pinus cedrus*). From *Plantae selectae*, vol. 6: t. 60 (1760), by C. J. Trew and G. D. Ehret. Artist G. D. Ehret. Image from the Biodiversity Heritage Library. Digitized by Missouri Botanical Garden. www.biodiversitylibrary.org.

PAGE 222 ❦ Cat in a Persea Tree, from *Book of Dead of Ani Sheet* 10. © Trustees of the British Museum.

PAGE 224 ❦ Inside Akiba Shrine, Utagawa Hiroshige, 1857. Courtesy of the Brooklyn Museum – CC BY 3.0.

PAGE 226 ❦ The fruits and leaves of *Dodonaea viscosa*, known as *akeake* in Te Reo Māori. From *Indigenous Flowers of the Hawaiian Islands*, t. 39 (1885), by Francis Sinclair. Artist I. Sinclair. Image from the Biodiversity Heritage Library. Digitized by Missouri Botanical Garden www.biodiversitylibrary.org

PAGE 230 ❦ A New Zealand Hebe (*Veronica speciosa*), known as *napuka* in Te Reo Māori. From *Favourite flowers of garden and greenhouse* Vol 3 (1897), by Edward Step. Image from the Biodiversity Heritage Library. Digitized by Missouri Botanical Garden, Peter H. Raven Library. www.biodiversitylibrary.org

PAGE 254 ❦ A species of sunflower (*Helianthus giganteus*) from North America. From *Botanica in originali* (1733), by Johann Hieronymus Kniphof. Photograph by Matthew Hall.

PAGE 280 ❦ The black mulberry (*Morus nigra*). From *De Historia Stirpium Commentarii Insignes* t. 522 (1542), by Leonhart Fuchs. Image(s) courtesy History of Science Collections, University of Oklahoma Libraries; copyright the Board of Regents of the University of Oklahoma.

PAGE 294 ❦ *Harakeke*, the New Zealand flax (*Phormium tenax*). From *Les Liliacées*, vol. 8: t. 448 (1805-1816), by P.J. Redouté. Image from the Biodiversity Heritage Library. Digitized by Missouri Botanical Garden, Peter H. Raven Library. www.biodiversitylibrary.org

ACKNOWLEDGMENTS

THE INSPIRATION FOR THIS WORK came from Val Plumwood. In her later works, Val called for environmental philosophy to converge with literature and poetry, to help create a culture of stories to reanimate the world. Thinking about Val's call from a botanical perspective, myths were an obvious source for stories about the interactions between humans and the mindful plant life we live among.

Seven years in the making, this manuscript has been through many iterations. What started as purely a collection of interesting mythological excerpts for a nonacademic audience, became part art book with the addition of stunning imagery, and then finally an even deeper hybrid, with the addition of some scholarly commentary. Along with Val's initial inspiration, I'd like to thank Graeme Gibson for his inspirational book *The Bedside Book of Birds*, which accounted for the decision to include imagery. I'd also like to thank the Metropolitan Museum of Art, Missouri Botanical Garden via the Biodiversity Heritage Library, the National Gallery of Art, University of Oklahoma Libraries, The Harvard Art Museum, Brooklyn Museum, the Yale Center for British Art, SLUB Dresden, and the Victoria and Albert Museum for the provision of accessible image databases for scholarly work, and for their kind permissions to use the imagery featured here. Thanks also to the publishers and rights holders who have allowed extended excerpts to be included in this work.

Over the course of seven years, I owe a debt to half a dozen publishers that have taken time to consider the manuscript, and to the dozen or so anonymous reviewers that have read and commented on the manuscript in that time. Your feedback has been invaluable in shaping the work, and I am very grateful for it. I'd also like to thank Freya Mathews for early encouragement and the late Nancy Ellegate at SUNY Press, who first saw the potential of this work, and after a circuitous journey, encouraged me to submit the manuscript to SUNY Press. Once the manuscript had found its natural home at SUNY, Christopher Ahn and his editorial team continued the support provided by Nancy, and Chris's guidance in particular has been invaluable. Thank you for your patience Chris! Feedback from two more anonymous referees added greater coverage and academic rigor. Thanks also to Ryan Morris at SUNY Press for invaluable production support and to Aimee Harrison for her wonderful book design.

The final, and deepest thanks go to my family. To my wife, Magnolia, for reading numerous drafts and for cajoling me to add more of myself at each step. To my eldest daughter, Iris, for inspiring me to keep going, and to my youngest daughter, Hebe, whose arrival provided ample encouragement to submit the manuscript on time.

Throughout
the inhabited
world, in all times
and under every
circumstance, the myths
of man have flourished;
and they have been the living
inspiration of whatever else may
have appeared out of the activities
of the human body and mind.
—Joseph Campbell,
The Hero with a Thousand Faces

INTRODUCTION

A Botanical Mythology

THE EARTH WOBBLES UNDERFOOT, the topsoil is broken and the tip of a small, gleaming sword pierces the surface. Between the circle of feet, the sword rises up, revealing a skeleton arm, then a whole skeletal body. The leader of the men looks astonished, but a few seconds later, the same thing happens—a wobble of earth and a skeletal warrior emerges. A second later, another emerges, then another, and another, until a whole army of skeletons has sprung up. The skeleton army attacks the gathered men and an epic battle is waged until until Jason escapes by jumping into the sea.

This is the memorable skeleton army scene from the 1963 film *Jason and the Argonauts*, a classic Hollywood movie based upon the mythological poem *The Argonautica*, written by Apollonius in third-century-BC Rhodes.[1] The cinematic version of Jason and the Argonauts utterly transfixed me as a child, almost twenty-five years after its release. Like the hero Jason, my eight-year-old self couldn't contain his amazement on seeing the skeleton army erupt from beneath the earth. In the few minutes that it took for that one scene to play out, the imaginative and inspirational nature of myth had taken hold of me. The idea of the skeleton men springing from the teeth of the Hydra that were sown in the earth by King Aeëtes (who was desperate to get his hands on the golden fleece) was almost hypnotic in its power. This was my introduction to the powerful world of myth.

In English-speaking nations, our understanding of this world is inextricably linked to ancient Greece. The term myth stems from *mythos,* the Greek for *word.* In his *Poetics,* Aristotle was the first to employ *mythos* in the sense of plot or "the organization of words and actions of a drama into a sequence of narrative components."[2] This is the idea of myth as "story." Myth as story is still the sense that prevails today. In his seminal work, *Mythography,* William Doty charts the linguistic development of myth and mythology:

> *Mythos*—"word" or "story"—could be combined with an equivalent Greek noun for "word," namely *logos* (related to the verb legein, "to speak"). The result: mythologia (English: mythology), literally "words

concerning words." However, historically, apart from its place in mytho-logia, *logos* gained the sense of referring to words comprising doctrine or theory, as opposed to *mythos* for words having an ornamental or fictional, narrative function.[3]

When Greek rational thought, from Plato onward, began to contrast rationality (*logos*) with the mythological (*mythos*) thinking of Homer and other epic storytellers, myth fell down on the side of the ornamental and the fictional.[4] Likewise, the distinction between myth and history saw myth fall on the side of the purely imaginative. Such distinctions surface in the work of scholars of myth. For the anthropologist E. B. Tylor, myths were primitive explanations of the world at odds with modern science—primitive ideas that were to be abandoned in favor of logic and scientific method.[5] In cultures such as ours, where science is the dominant form of knowledge, such a view of myth is well established. For Tylor, as for many others, the word *myth* is synonymous with *untrue*.

We are not bound to viewing myths as entertaining but ultimately misguided interpretations of the world. Scholars have questioned the opposition between rationality and mythology, and the understanding of myth as *untrue*. William Doty acknowledges the *fictional* or narrative character of myth, but emphasizes the important social role of myth, "that of modelling possible personal roles and concepts of the self."[6] A positive view of myth is that it is fundamental to cultures, and has a significant role in shaping religious beliefs, philosophies and worldviews. In her short history of myth, Karen Armstrong argues that myth is not an inferior mode of thought to our cherished logic and reason, but is:

> a game that transfigures our fragmented, tragic world, and helps us to glimpse new possibilities by asking "what if?"—a question which has provoked some of our most important discoveries in philosophy, science and technology.[7]

For Armstrong, the source of mythology is the imagination, and the role of myth is to extend the scope of human beings, using stories to force us beyond our own experience, to create a sense of belonging and to show us how to behave in the world. Myths then are "stories about something significant."[8]

Such a view of myth draws heavily on mythographers such as Mircea Eliade and Joseph Campbell. Both have argued that myth has played a central role in the development of human society and culture, the description and presentation of human knowledge, the construction of relationships, and ulti-mately our identity as human beings.[9]

> It wouldn't be too much to say that myth is the secret opening through which the inexhaustible energies of the cosmos pour into human cultural manifestation. Religions, philosophies, arts, the social forms of primitive and historic man, prime discoveries in science and technology, the very dreams that blister sleep, boil up from the basic, magic ring of myth.[10]

All this was expertly and succinctly summarized by Claude Levi-Strauss, in his seminal study of mythology—"With myth," he wrote, "everything becomes possible."[11]

A HUMAN-CENTERED WORLD

As someone with an interest in the perspectives of nonhuman beings, my own take on mythology is that many myths, and their accompanying scholarly inter-pretations, are human centered. Regardless of the theoretical interpretation of what myth *is,* from William Robertson-Smith's view of myth as ritual, to Kenneth Burke's understanding of myth as metaphysics, many myths are framed and presented so that the human heroics (or at least gods in human form) are at the center of the narrative. A number of key figures in the history of mythography, including Campbell, Otto Rank, and Lord Raglan, have explicitly foregrounded the human hero as the principal subject of mythology. Myths such as the fall of Phaethon, Odysseus's journey home after the fall of Troy, and the account of the enlightenment of the Buddha are all treated as myths of the human hero.

The power of these human hero stories has partly enabled these myths to persist into the modern day. Human hero myths fascinate humankind and they permeate and penetrate all aspects of human life and society. This influence extends into some of the most powerful contemporary cultural forces, advertising and mass media.[12] Where I was entranced by Jason and the Golden Fleece, young children of the past decade have been introduced to the human-hero mythology through hugely popular children's literature and cinema, such as *Harry Potter* or *The Lord of the Rings.* The human hero on a voyage of self-discovery is also the basis of the modern novel (have you read *Don Quixote, Robinson Crusoe, Great Expectations?*), of a large proportion of the output of Hollywood movie studios (remember *Casablanca, Ferris Bueller's Day Off, Thelma and Louise,* or *The Life*

Aquatic?) the resurgent television drama (just watch *Breaking Bad*) and (sort of) popular music (listen to Björk, Nick Cave, Daedalus).

It would not be too much an exaggeration to say that the myth of the human hero is our primary narrative. In such myths, the dynamic human beings are placed front and center in their quests, often independent of nature and elevated above it. The other elements of the natural world, when they appear, are secondary to the human; they are largely the backdrop against which the human narrative takes place. The Ionian Sea is the setting for Odysseus's journey to Ogygia, the Bodhi tree is the place of the Buddha's enlightenment, and the sun is the backdrop for Phaethon's deadly fall. The natural world is there, but it is eclipsed by the human (largely male) hero. This observation is in line with eco-feminist accounts of human culture, in which the human reason, connected with the male, is valued over a nonreasoning nature, connected strongly with the female.[13]

Myths in which humans are first and foremost have undoubtedly influenced our relationship with the natural world. We live in a world that has been made to serve humanity at every turn, a world in which our primary interaction with nature is to treat it as a set of resources, a backdrop, an instrument for the satisfaction of human desires. The human-centered view of mythology neatly mirrors, and no doubt informs such a human-centered, or anthropocentric, world. The anthropocentric worldview found in hero mythology underpins a situation in which human beings and their welfare are the sole focus of modern society and are regarded as the primary source of value in the world.

A worldview in which everything revolves around human wants and desires ultimately rests on the idea that we should only show moral concern for human beings. Humans alone possess the "advanced" characteristics deemed worthy of respect, such as intelligence, language, and reason. In this view, the natural world is not a focus of ethical behaviors. Instead, it is relegated to a footnote, cast as a passive collection of resources for us to use as we wish in the ongoing drama of human society. Depicted as "lacking" these human attributes, animals sit significantly lower in a hierarchy of nature. Plants are at the bottom of the hierarchy; passive, insensitive, and unthinking.

The problem with this anthropocentrism, as many leading environmental thinkers have pointed out for decades, is that it lurks behind our society's rampant disregard for nature and the widespread and ongoing degradation of natural ecosystems.[14] Human society is obsessed with the mythology of the heroic, superior human hero that uses the natural world as his or her dramatic backdrop. A thousand car advertisements will testify to the fact that in our understanding of both mythology and modern life, the human hero is active and dynamic, the rest of the natural world is backgrounded and largely passive, available for use as we heroic, "superior" humans see fit. Not only is this a planet now at risk of ecological collapse, it is a duller, less vibrant world in which the presence, abilities, and needs of other species are obscured behind a cloud of human exceptionalism.

PLANTS IN THE ACTIVE VOICE

For the great environmental thinker Val Plumwood, countering this anthropocentric position requires us to undertake the immense challenge of "(re)situating humans in ecological terms and non-humans in ethical terms."[15] A significant contribution to this challenge requires a leveling of the playing field between humans and nonhumans so that human beings are not the only heroes in the mythology that underpins contemporary culture. Redressing this balance is itself a mammoth task. It requires a radical change in the way in which humans understand nature, remaking nature in the "active voice"—that is, recognizing and theorizing nature and her species as volitional, purposeful, and mindful.[16] To do so, other natural species need to be the focus of dynamic and heroic stories. As the basis of the natural world, we need plants in particular to step out of the shadows of human instrumentalism. Plants urgently need to become the focus of our modern myths.

There are many accounts of plants that present them primarily as instruments or objects for human use or pleasure. Each year, hundreds of books are written about plants that present their stories almost entirely from the human perspective—whole volumes dedicated to the horticultural beauties, delectable garden delicacies, or plants as the objects of philosophical musings.[17] When plants in mythology are written about, those plants are often presented as symbols for human characteristics, or constituents of human dramas.[18] As chapter 1 explores, such a view of plants is ingrained in the European philosophical and religious heritage.

In order for nonhuman species to become heroes rather than supporting acts, we require a culture of stories that allows plants, from the garden anemone to the majestic kauri tree, to impress their mindful, active, unique natures upon our human imagination. We need a body of narratives in which the plant in question is presented as much as possible as an agent—an "independent centre of value, and an originator of projects that demand my respect."[19] We need to tear ourselves away from looking longingly at our own human reflections (like Narcissus) and pay proper heed to the plants that make possible our life on this Earth.

Fortunately, to build up such a body of stories we need not start from scratch; we need not begin penning a new *Odyssey* in which the hero is an animal or plant rather than a human being. In the world's mythological canon, although mythography, and common understanding, has privileged the human, there are in fact a multitude of stories and tales in which the human hero and human drama is not the be all and end all.

In the course of writing my first book, *Plants as Persons*, I became aware of a myriad of myths and stories that presented the most unlikely of subjects—plants—as more than just the silent servants of humanity. In the texts and traditions of cultures from all over the world, I stumbled across myths that demonstrate, explore, and present plants (and through them the wider natural world) as sensitive, communicative, and intelligent. This is a far cry from the numerous presentations of plants in mythology as predominantly symbols of human truths.[20]

AGAINST ANTHROPOMORPHISM

The philosopher Michael Marder has asserted that using human terms and concepts, such as intelligence and sensitivity, to describe plants is a form of anthropomorphism.[21] Marder writes:

> Taken together, the projections of the human onto the plant and of the plant onto the world are tantamount to a metaphysical transposition of the human onto nature as such, the transposition, where the domesticated and homologous fragments of vegetal life are used as the means in the narcissistic self-recognition of the human in the environment. (Let us recall, in this context, that the concept of narcissism is, itself, derived from the name of a mythical character—Narcissus—that was bestowed upon a flower, thereby completing the enchanted circle of the anthropomorphization of plants and the vegetalization of the world.)[22]

Marder's accusations of anthropomorphic self-projection are a familiar attempt at "delegitimating any new or old animating sensibility."[23] The concept of anthropomorphism is itself ambiguous, bearing both the concepts of attributing to nonhumans characteristics that humans have, and attributing to nonhumans characteristics *only* humans have. The first sense assumes that there is no overlap between the characteristics of humans and nonhumans and the second sense is simply question begging.[24] Both ignore the continuities between human and nonhuman life and rest upon an understanding of human and plant natures as hyperseparated.

This charge of anthropomorphism is often thrown around when the continuities of human and more-than-human lives are highlighted, and as Plumwood astutely points out:

> That has become its major function now, to bully people out of "thinking differently." It is such a highly abused concept, one often used carelessly and uncritically to allow us to avoid the hard work of scrutinising or revealing our assumptions, that there is a good case for dropping the term completely.[25]

In addition to its carelessness, Marder's view also conveniently ignores those cultures that have constructed ethical relationships of care and respect for plants from acknowledging continuities between the human and the more-than-human worlds. For example, in many Indigenous cultures, relationships between humans and more-than-humans rest upon a recognition of the continuities between human, animal, and plant life (explored in chapter 1) and this recognition of similarities and continuities is often expressed in human terms (what other terms do we have?).

The language of sentience and intelligence can be viewed, not as a self-projection, but, to adapt Voloshinov's famous line, as a *bridge* between two *types* of being, from human to the plant that is so obviously different in its outer and inner form.[26] This language is a form of empathy, employed in the service of building relationships of care and kinship. In essence it provides the *sameness* required for flourishing relationships in the face of obvious morphological difference.[27] As chapter 1 will explore, a language of sameness and continuity is also the natural expression of kinship relationships in which the "mutuality of being"[28] is the primary, defining, characteristic.

THE IMAGINATION OF PLANTS

The imaginative power of myth allows us to move beyond our mundane experience and guides us to a live a more fulfilling, richer life. Myths open up the possibility of new relationships, and new ways of living. They provide us with "aspirations toward becoming something other than what we are" and "ways as to imagining new possibilities as to who we are."[29] The *Imagination of Plants* is a collection of the rich botanical mythology that aims to use myths to push beyond the boundaries of our ordinary experience of plants, and to challenge the way in which we humans understand and relate to the plant kingdom and the wider natural world. This in turn is a challenge to human identity, most particularly our innate sense of superiority.

Like much of this work, this approach also takes direct inspiration from Val Plumwood, who championed a project of reanimating the world, and with it situating human beings as members of the ecological community.[30] In this vital eco-political quest to reanimate the world, Val recognized that the power of story is key:

> We are in desperate need of stories that create much greater transparency of these [ecological] relationships in our day-to-day lives. We must once again become a culture of stories—stories that link our lives with the Great Life which some call Gaia, but all should call by names of their own devising. This is the real meaning of ecological literacy, to have stories that speak of the culture/nature boundary and of where the two cultures meet. Instead we have one discourse about the domain of culture (us) and another discourse, formulated in an especially detached and distant way, about the domain of nature (them). Our conviction that "we" live in culture and "they" live in nature is so strong that all that is left is a passionate story about consciousness, history and freedom—about us—and another story about fiercely uninvolved causation and clockwork—a story about them.[31]

Myths are an obvious source for stories about the interactions between humans and the mindful plant life we live among and depend upon. The *Imagination of Plants* discusses and presents extended mythological excerpts (what can also be called analecta, from the Greek *analektos*—gathered together) about plants in which plants are often active subjects in the stories in question. These excerpts are themselves from a subset of the works referred to in each chapter commentary, selected for their diversity and quality. The commentary itself critiques and finds common threads between these stories, with a view to redressing the bias of the human hero and the human separation from nature.

This, unashamedly, is a work of comparative mythology and is in line with what Doty terms "comparative thematic elucidation," to describe a "type of freely associative study that consists of tracking motifs and pattern similarities no matter where they originally occur."[32] Such cross-cultural comparative analyses "can be misleading if they are considered as providing genetic explanations" but can be of great value when "they are used to establish a projective matrix of possible realizations of a particular theme" and in particular what is unique about each.[33]

The Imagination of Plants should be thought of as an attempt to demonstrate the existence of a series of ideas, motifs, and themes in myths concerning plants. The excerpted selections are an attempt to exemplify, rather than define or verify.[34] As such, I have no interest in providing "genetic" explanations of these myths or of reading back into these texts any presumed significance for the cultures at hand. Moreover, this work is not an unthinking collation of different accounts, with no regard to the historical and cultural context, in the manner of James Frazer's *The Golden Bough*. The historical and cultural context is considered, particularly that of the texts themselves, but this is not placed front and center (see Guide to the Texts for this material). The principal aim of this work is to reposition the human relationship with the plant kingdom and to foreground plants as much as possible. It looks to imagination and ideas inherent within mythology as the source inspiration for this endeavor.

This cross-cultural comparative analysis draws on a collection of myths from dozens of different cultures, from Aboriginal Australia to Zoroastrian Persia. It uses what can be thought of as classical mythological sources for ancient Greece such as Ovid's *Metamorphoses*, Virgil's *Aeneid*, Homer's *Odyssey*, as well as other influential epics and the stories they contain, such as *Gilgamesh* (Mesopotamia), *The Kalevala* (Finland), *The Mahābhārata* (India), and the *Kojiki* (Japan).

However, selections are also included from texts that may be thought of as more religious in nature. I have included excerpts from these texts as they are important sources for stories that concern our relationship with the botanical world. Texts used include the Hebrew Bible, the Zoroastrian *Bundahisn*, and the Indian texts *Rig Veda*, *Padma Purana*, and a number of the *Upaniṣads*. Another important source for botanical myths are mytho-historical texts such as the *Nihongi* of Japan, and the *Popol Vuh* of the Maya. Contemporary collections of oral stories are also important sources of myths from Indigenous cultures, including the Acoma of New Mexico, Māori of Aotearoa, and Aboriginal Australia. For the discussion of plant legends in chapter 4, nonmythological texts such as Pliny's *Natural History* and the *Travels of Sir John Mandeville* have been particularly useful.

In the seven years that this manuscript has been in development, my one constant intent has been to make this material as accessible as possible to the general reader. The selections provided are often, therefore, from translations of works already in the public domain, both to allow lengthy inclusions and to enable the reader to freely access the full text for further reference. Although providing non–English language texts in translation may result in the loss of some of the linguistic and poetic nuances, "the mythical value of the myth remains preserved, even through the worst translation."[35] Even so, many of the excerpts from classical Greek and Latin texts are provided from translations that have strived to literally interpret the original text, e.g., Miller's translation of Ovid's *Metamorphoses*. While more poetic or modern translations are available, a literal translation provides perhaps the best foundation for our task of understanding the portrayal of plant life in myth.

In order to help the reader to reflect, meditate, and muse upon their own relationship with the plant kingdom, the excerpts of botanical mythology from this wide range of texts have been curated into six thematic chapters—Roots, Gods, Metamorphosis, Legend, Sentience, and Violence. These six chapters, and the multiplicity of myths that they contain, are underpinned by two major themes, kinship and sentience. These themes appear again and again when examining the stories of plants across dozens of different cultures and crosscut the six thematic chapters in multiple instances.

The themes of kinship and sentience (themselves inseparable) are predominant in the botanical mythology from across the world and many of the myths presented feature both. The creation stories which are the subject of chapter 1, such as the Māori stories of Tāne and Rātā, tell of human-plant kinship through shared origins, plant sentience (chapter 5), and the fact that plants are capable of being subject to violent actions from human beings (chapter 6). Many of the creation myths from across the world contain descriptions of the sacred plant species that are featured in chapter 2.

I hope that readers will derive inspiration from the myths discussed and presented; inspiration for both questioning and reimagining their own relationship with the plant kingdom. While the introductions to each chapter will provide some context, critique, and guidance, the theorizing has been kept to a minimum. I have tried to present the myths simply as stories about plants— sometimes literal, sometimes symbolic, often allegorical. The structure of this work has also been designed to give readers the chance to read and reread the botanical myths for themselves. The myths are discussed under each theme and the majority of the myths discussed (but not all) are also presented as excerpts.

Only by reading the myths first hand will the reader be able to form their own opinions of these portrayals of plant life. The textual excerpts then are fundamental to understanding the thematic presentation and critical discussion of these botanical myths. The Guide to the Texts is included at the end of the book for those who wish for more detail about the sources. A large number of the excerpts are also accompanied by striking images of botanical or religious art, which feature scenes from the botanical myths and associated characters, such as gods, animals or the plants in question. Taken together, the hope is that these beautiful pieces of literature and art will inspire each reader to take their own approach to the myths at hand and to use them to reflect on their understanding of the plants that form part of their own lives.

I offer this material for inspiration in the spirt of Val Plumwood, who, more than a decade ago, urged her readers to

> free up your mind, and make your own contributions to the project of disrupting reductionism and mechanism. Help us re-imagine the world in richer terms that will allow us to find ourselves in dialogue with and limited by other species' needs, other kinds of minds. I'm not going to try to tell you how to do it. There are many ways to do it. But I hope I have convinced you that this is not a dilettante project. The struggle to think differently, to remake our reductionist culture, is a basic survival project in our present context. I hope you will join it.[36]

Although the ancient heathens did appropriate the first invention of the knowledge of herbs, and so consequently of physic, some unto Chiron the centaur, and other unto Apollo or Aesculapius his son; yet we that are Christians have out of a better school learned, that God, the Creator of Heaven and Earth, at the beginning when he created Adam, inspired him with the knowledge of all natural things: for, as he was able to give names to all the living creatures, according to their several natures; so no doubt but he had also the knowledge, both what herbs and fruits were fit, either for meat or medicine, for use or for delight.
—John Parkinson, *Paradisi in Sole*

ROOTS

PUBLISHED IN ENGLAND IN 1629, in the reign of Charles I, John Parkinson's *Paradisi in Sole Paradisus Terrestris* is one of the foundations of modern literature on gardens, gardening, and the cultivation of garden plants. A Yorkshireman by birth, like many of his countrymen Parkinson was a pragmatist. He sought to impart a practical knowledge to his readers, and in the course of doing so he challenged long-held beliefs about the nature of plants found in earlier English gardening texts. Where earlier authors had asserted the effectiveness of practices such as soaking seeds in different colored dyes to affect the color of petals, or perfuming fruit trees so that the produce would take on the scent, Parkinson regarded such methods as "idle tales and fancies."[1] As the prefacing address to his readers attests, the knowledge of pre-Christian Europe was also dismissed, the author instead founding his view of the purpose and nature of plants on the Book of Genesis and the Garden of Eden.

In Parkinson's archetypal Eden, human beings are front and center. Eden is planted as a home for Adam, and is filled with plants that are useful for food or medicine or which give aesthetic pleasure. Although this doesn't seem so strange, what I find so interesting about the myth of Eden is that in the story of their own creation, plants are merely a backdrop for human existence. Their entire lives are turned over to the service of humankind. So, on the authority of the myth of creation in Genesis, Parkinson tells his readers that "God made the whole world, and all the creatures therein for man, so he may use all things as well of pleasure as of necessity."[2] In this understanding of the natural world, plants have been created to be resources—food, shelter, medicine, clothing—or to give us pleasure, providing a beautiful home for the human beings in Eden. There is no sense at all that plants live for themselves.

Plants can be cast in this way in Genesis because of the text's take on human superiority. Although the whole earth was created by God, humans are placed above and separate from the other creatures in God's creation, elevated from the rest by being the only creatures made in the image of God.[3] In later philosophical writings, such as those of St. Thomas Aquinas and John Locke, this separation is maintained by decreeing humans to be the only beings bestowed with intelligence and reason.[4] In a famous article, the scholar Lynn White Jr. has argued that this Judeo-Christian separation from the natural world is at the heart of our current ecological crisis.[5]

Although such readings are being increasingly rejected by the Church,[6] if you see plants as inferior creatures that are here to provide you with all that you need, I would argue that a Parkinson-like reading of the myth of creation in Genesis has shaped your understanding of plants. If the plants in your garden are there either for their looks or for their tastiness (the idea of anything else is bordering on the sacrilegious to most gardeners), you are not only in good company with John Parkinson, but also with the authors of most books about plants—texts in which plants are considered almost entirely from a human perspective, praising their beauty or usefulness. This view of plants may seem pretty normal and perfectly reasonable, but in the context of the world's mythology, such an understanding of plants is not even half the story. The creation myths of the world are our first stop on a search for a different perspective.

KINSHIP: A MUTUALITY OF BEING

Mircea Eliade's contention that all myths are ultimately myths of creation[7] is still debated by scholars, but it is generally agreed that creation myths (cosmogonic myths) are of enormous power, and have a significant effect on the human understanding of the world. Across the world, from South American pre-Colombian peoples, to Aryan Indians, pagans, and Aboriginal Australians, cosmogonic myths are a significant source of power that, whenever these stories are told, is renewed and made manifest.[8] The retelling and recycling of cosmogonic myths has a significant influence on human cultures and the human perceptions of and behaviors toward plants.

From a plant point of view, the key to understanding such myths, and the way in which they portray connections between human beings and plants, is the concept of *kinship*. The anthropologist Marcel Sahlins has produced one of the most comprehensive reviews of the notion of kinship in recent times. For Sahlins, kinship is essentially a "mutuality of being," a concept which applies to interpersonal relationships that are cosanguineal (with shared blood) and affinal (through marriage). This mutuality of being comes in different forms, but generally

kinsmen are persons who belong to one another, who are members of
one another, who are co-present in each other, whose lives are joined
and interdependent.[9]

Sahlins is at pains to emphasize that kinship can be brought into being
through actions, behaviors, and rituals involving persons, but at the same
time recognizes that kinship also emerges through shared ancestry and substance (e.g., shared blood).[10] Here he references Aristotle's *Nicomachean Ethics*,
the foundational classical text, which grounds kinship in shared origins and
shared substance,

> for the self-sameness of their relation to *those* produces the same with
> each other (hence the way people say "same blood," "same root," and
> things like that). They are, then, the same entity in a way, even though
> in discrete subjects.... The belonging to each other of cousins and other
> relatives derives from these, since it exists by virtue of their being of the
> same origins, but some of these belong more closely while others are
> more distant, depending on whether the ancestral common sources are
> near or further off.[11]

In many cultures, such relationships have a significant effect on the notion
of self, and "relatives are perceived as intrinsic to the self."[12] Kin relationships,
then, are characterized "in terms of a feeling of 'love, longing, care' for a person,
a mental activity of 'caring for, loving' another person, or a moral position of
being 'unitary, solidary, amicable' with someone."[13]

KINSHIP: COMMON ORIGINS

In many myths, counter to the idea that John Parkinson found in the Book
of Genesis, that humans are above plants and wield dominion over them,
plants and humans are depicted on the same level, as kin sharing common
ancestry. The expressions of human-plant kinship found in creation myths
are as many and varied as the cultures that tell these stories. But, from examining the depiction of plants in creation myths across many cultures, there are
three common ways in which human kinship with plants is demonstrated in
botanical mythology.

The first of these (as Aristotle noted) is grounded in a common origin.[14]
Many mythological traditions tell us that if you look back far enough you'll

find a common ancestor between human beings and plants. Unfortunately for those who think of myths as "primitive" (such as E. B. Tylor), modern phylogenetics tells us just the same. We both emerged from some common single-celled ancestor 1.6 billion years ago. Although we may not live this truth, it is a scientific fact that you and the rice plant (*Oryza sativa*) are kin.

Rather than a single cell, in many myths the common ancestor is godly in form and nature. In the telling of an Icelandic creation myth in the *Gylfaginning* (a text of Snorri Sturluson's *The Prose Edda*), the common ancestor of humans, animals, and plants is the god Ymir, whose death creates the entire natural world. Ymir becomes our common ancestor when his decaying body makes the earth and sky, the seas and the trees:

> Of Ymir's flesh the earth was fashioned.
> And of his sweat the sea;
> Crags of his bones, trees of his hair.
> And of his skull the sky.[15]

The death of Ymir creates the lineages of all the plants and animals. In the myth of Genesis, it is the Hebrew God that creates the plants and animals, and there is a subtle difference between these two modes of creation. If you read through the whole excerpt from the *Gylfaginning* in the following pages, you might notice that the distancing of plants and humans is absent. Whereas in the account of creation in Genesis only man is made in the image of God, in this pagan myth, the very body of the god Ymir creates the lineages of both human beings and plants. The connection is emphasized and foregrounded. Certainly, one relation isn't elevated above (or enslaved to) another.

This creation of the world from the death of a common ancestor, and the shared genealogy of humans and plants, isn't just confined to a small part of northern Europe. The death of a divine being whose body creates the world is a feature of many Indo-European creation stories, also occurring in the Indian Vedic texts that are the spiritual foundations of Hinduism.[16] In the excerpts at the end of this chapter it is interesting to compare the similarities and differences of each.

The *Rig Veda*, one the oldest texts of the Indo-European tradition, tells of the sacrifice of another common ancestor, the divine being Puruṣa (meaning person or man). After being sacrificed by the gods (he must have done something pretty bad), Puruṣa's dismembered body becomes the various castes of Hindu society and also creates the natural world. Through Puruṣa, the genealogies of plants and animals, including humans, become entwined. As in the

Vǫluspá, not only do they share an abstract "creator," they share the same substance, a marker of the mutuality of being that characterizes kinship.

This myth is echoed at the beginning of the *Bṛhadāraṇyaka Upaniṣad* in which the disembodiment of a sacrificial horse, rather than a human, creates the physical world:

> Heaven is the back, the sky the belly, the earth the chest, the quarters the two sides, the intermediate quarters the ribs, the members the seasons, the joints the months and half-months, the feet days and nights, the bones the stars, the flesh the clouds. The half-digested food is the sand, the rivers the bowels, the liver and the lungs the mountains, the hairs the herbs and trees.[17]

The story of plants and animals becoming kin through the death of a common ancestor also occurs in the *Bundahisn*, the Persian encyclopedic collection of cosmogonical and cosmological stories. The first being is not a horse this time, but an ox—a primeval ox, which has various names, including *Evagdad*.[18] The *Bundahisn* relates that the ox is a hermaphrodite, having both semen and milk. When placed in the heavenly spheres it is attacked and dies. From the marrow of the dying ox, grains and grapevines are brought into existence, along with numerous other animal species. Like Ymir and Puruṣa, the ox is the start of the separate lineages of plants and animals, but unlike the preceding two myths, human beings are absent. Humans are not created from the death of the primeval ox, but from the death of the primordial human Gâyômard, who is described as being "the first man." It could be argued that this is very similar to the account of creation in Genesis, but unlike for Adam and Eve, the kinship between human beings and plants is still strongly evident. Following the death of Gâyômard and the subsequent planting of a seed in the earth, the first human beings, Matro and Matroyao, are born from, and in the form of, plants. Here, the plants themselves are the common ancestor, as the myth tells us:

> And in forty years, with the shape of a one-stemmed Rivas-plant, and the fifteen years of its fifteen leaves, Matro [Mashye] and Matroyao [Mashyane] grew up from the earth.[19]

These wonderful tales about our common ancestry with the plants, animals, rocks, and other parts of the Earth are not restricted to Indo-European sources. One of my favorite accounts of the common ancestor is the popular Chinese creation myth of Pangu, which first appeared in Xu Zheng's *San Wu Liji* in the third century CE. Pangu is the world creator, whose death forms the world and all its elements. In a later account by Werner, as Pangu died,

[h]is body began to change and his breath became the wind and clouds and his voice the thunder. His eyes became the sun and moon and his limbs the mountains and valleys. His blood formed the rivers and seas and his hair became the plants and trees. His bones changed to become rocks and his sweat became the rain and dew. The parasites on his body became all the animals and humans.[20]

In the myth of Pangu, plants and humans are depicted as sharing a common ancestral origin. What is most interesting about this account is that human beings are not elevated above the rest of creation. They are created from the parasites on the body of Pangu (perhaps quite aptly), and are made from exactly the same substance as the rest of the animal kingdom. Although plants are created from the hair, they still share Pangu as a common ancestor.

There must be something about human hair and plants, because this is a connection that is found in a number of other mythological traditions. The link appears in a Native American Okanogan myth which relates that "[t]he earth was once a human being. Old One made her out of a woman.... Earth is alive yet, but she has been changed. The soil is her flesh, the rocks are her bones, the wind is her breath, trees and grass are her hair."[21] It is also found in story of Aino, in the collection of old Karelian myths collated by Elias Lönnrot into the Finnish epic poem *The Kalevala*.[22] I am not sure where this connection comes from, but it is unlikely that both Native Americans and ancient Karelians both came into contact with the hair-like American native species *Tillandsia usneoides*.

KINSHIP: DIVINE (COMMON) ORIGINS

The second way in which myths show human-plant kinship again features a common relative, but these myths do not involve the death of this ancestor. Instead, common ancestry is shown by the creation of living beings from the living body of the creator god. This motif again combines the shared substance and the shared ancestry that are strong markers of kinship relationships. A good example of this type of kinship appears in the Indian sacred text the *Vishnu Purana*. In this text there is a passage that describes plants once more being made from human hair, this time from the body hair of Brahma, the Hindu god of creation:

> Sheep were created from his breast, horses, elephants and other animals from his feet and "from the hairs of his body sprang herbs, roots, and fruits."[23]

Similarly, in Egypt, both plants and human beings can trace their origins to the gods Rem (a fish god who fertilizes the land with tears) and Nu (the personification of the primordial waters). A passage from the *History of Creation* (Papyrus BM 10,188) relates that the tears of the God Nu created both vegetation and the men and women of the earth.

> Plants and creeping things sprang up from the god Rem, through the tears which I [Nu] let fall. I cried out and men and women came into existence. Then I bestowed upon my Eye the cobra of fire, and it was angry with me when another power (the Moon) came and grew up in its place; its vigorous power fell on the plants, on the plants which I had placed there, and it set order among them.[24]

This is one of a number of instances in Egyptian mythology in which the divine bodily fluids are transformed into plants.[25]

KINSHIP: THE FIRST GARDENERS

The third way in which myths can establish human-plant kinship involves what I like to think of as a spot of cosmic gardening. I am referring here to stories in which the ancestors (often divine beings) bring plants to the land in the time of first creation, often planting the vegetation throughout the ancestral lands. These beings are the common ancestors of the plants, animals, and humans, but what is interesting about these myths is that they combine shared ancestry with a depiction of plants as sentient beings. Plants are presented in such a way that they are not solely for humans to use as they wish.

In the opening of the *Kalevala*, in the time of the creation, it is the mythical hero Pellervoinen who sows all the vegetation on Earth, including numerous tree species such as juniper, oak, and pine.[26] Pellervoinen sows seeds of plants that are animate and fully alive—on many occasions engaging the main human characters in conversation. Through Pellervoinen, humans and plants have a shared genealogy, and can both trace their origins back to his actions.

Although there is little linguistic or cultural connection between Karelia and Mayan South America, the same idea also occurs in Quiche Mayan mythology. According to the *Popol Vuh*, four ancestral animals brought the corn plant (yellow and white) to the creators of the Mayan people. The forefathers not only provided this corn for the first humans to eat, but they also used corn to make the flesh and blood of humankind. As well as having common ancestors in the four animals, humans and plants share the same substance.[27] These continuities form the foundation of the human-plant relationships in the *Popol Vuh*.

In North America, the origin myth of the pueblo of Acoma tells of two ancestral humans born underground in the dark. Living in the dark, they grew slowly and were impatient to get out into the light. One day they were given a present of pine tree seeds from a woman called Tsichtinako:

> Tsichtinako instructed them, "You will find the seeds of four kinds of pine trees, lă'khok, gēi'etsu (dyai'its), wanūka, and lă'nye, in your baskets. You are to plant these seeds and will use the trees to get up into the light."[28]

The humans plant these seeds and then, once sprouted, these plants share the same origins and qualities as the humans. They are born in the earth and grow slowly in the dark. Like the ancestral humans, they are also active in the creation of the world—the *lă'nye* tree grows up to poke a hole in the earth and let in the light. Once again, the human and the plants of the Acoma lands can trace their origins to the actions of Tsichtinako. Humans and plants have a shared ancestry. Humans and plants are kin.

In a Zuni myth, the hallucinogenic plant *Datura innoxia* also has a familial connection with human beings, sharing both its origins and name with the first humans A'neglakya and A'neglakyatsi'tsa.[29]

> In the olden time a boy and a girl, brother and sister (the boy's name was A'neglakya and the girl's name A'neglakyatsi'tsa), lived in the interior of the Earth, but they often came to the outer world and walked about a great deal, observing everything they saw and heard and repeating all to their mother. This constant talking did not please the Divine Ones (twin sons of the Sun Father) ... so the Divine Ones caused the brother and sister to disappear into the Earth forever. Flowers sprang up at the spot where the two descended—flowers exactly like those that they wore on each side of their heads when visiting the Earth. The Divine Ones called the plant a'neglakya (*Datura innoxia*) after the boy's name. The original plant has many children scattered over the Earth.[30]

The fact that the *Datura* shares a name with the first humans adds strength to the ties created through shared ancestry.

Similar stories of ancestors bringing the first plants to the land are common in the mythology of Australian Aboriginal peoples. The stories of the Dreaming beings are the foundation for a basic sense of kinship between human beings and plants and animals. The Dreaming beings are ancestors, who were born from the earth and who created and shaped the Australian landscape. These beings and their actions create an immanent presence in the landscape, and one that imparts autonomy to all beings. This presence persists for as long as the landscape exists, for in the words of eminent Australian anthropologist Ted Stanner, the Dreaming was not just a time past, but an ever-present "everywhen."[31]

The majority of Dreaming stories largely concern animal species, but there are a significant number also involving plants. Many of these concern the metamorphosis of Dreaming beings from an initial humanity into a plant form. These once-human beings transformed into the Iga tree, the Pandanus, and the lilies. As in the tales of Ymir and Pangu, both the human and the plant share common ancestry in the Dreaming beings. But the plants that can trace their origins to the actions of a Dreaming being are also regarded as autonomous. In a Ngarinman account of creation, from the Northern Territory, Australia, the plants are subject to their own Law (their own rules and purposes)—the "Plants started growing according to their own 'laws'—their own shape, size, habitat requirements, and 'behaviour.'"[32]

Being subject to their own Law does not mean that plants are not eaten. The creation stories tell how these plants were distributed by the actions of Dreaming beings who brought food to country. For Yanyuwa people, the cycads grow in their country as a result of the Tiger Shark who "threw the cycad nut everywhere, over long distances he threw it."[33] This traditionally important food for the Yanyuwa has its origins in the actions of the spirit ancestor. In Mayali country, Warramurrungundji brought important food plants such as *anbonde*, a banyan tree (*Ficus virens*), and *anguladji*, the spike rush (*Eleocharis dulcis*), whose rhizomes are a good food source for people and magpie geese.[34]

In these accounts, the Dreaming beings are not just the ancestors of human beings; they are ancestral to humans, animals, and plants. They create the kinship relationships between humans and nonhumans that are characteristic of Aboriginal cultures. As the final chapter explores, these kinship relationships are built as much on violence and killing as on a recognition of sentience.

Across the Tasman Sea in Aotearoa/New Zealand, the *whakapapa* (ancestry) of each forest tree can be traced back to the Māori forest ancestor (*atua*), Tāne. George Grey's version of the Māori creation myth tells of Tāne separating his parents Rangi-nui and Papa-tū-ā-nuku as the principal act of creation, the act that formed light and dark.[35] Rangi-nui and Papa-tū-ā-nuku's other children are the atua of the rest of the living world, including Tūmatauenga (humans), Tāwhirimātea (weather), and Tangaroa (sea, fish).

The shared *whakapapa* between forests and human beings is the foundation of strong kinship relationships between Māori and the forest.

In John White's *Ancient History of the Māori*, the story of Tāne goes farther. White includes a story of Tāne taking specimens of every tree species from heaven and planting them in the Earth. The trees bear the marks of such ancestry to this day.[36] This tale of bringing plants to the land has strong parallels with those from Aboriginal Australia and the Māori migration stories that celebrate the bringing of the staple crop *kumara* (*Ipomoea batatas*) in the canoes of the Māori ancestors, from Hawaiki, the mythical homeland, to Aotearoa.

In a similar fashion to Tāne's adventures in heaven, the eighth-century text of Japanese classical history, the *Nihongi* (also known as the *Nihon Shoki*), relates that when the character "Iso-takeru no Kami descended from Heaven, he took down with him the seeds of trees in great quantity."[37] Following his descent he planted the seeds all over the land of Korea, creating rich green hills in the process. But Iso-takeru no Kami saw the need for greater botanical riches, to rival those in "the land of the Han," that is, China:

> So he plucked out his beard and scattered it. Thereupon Cryptomerias were produced. Moreover, he plucked out the hairs of his breast, which became Thuyas. The hairs of his buttocks became Podocarpi.[38]

Here a shared ancestry, the creation from body parts and bringing of plants by ancestors overlap and interpenetrate. Iso-takeru no Kami created plant lineages through both bringing seeds from heaven and the sprinkling of hair from beards, breast, and buttock!

LIVING WITH KIN

Claude Lévi-Strauss wrote:

> On the one hand, a myth always refers to events alleged to have taken place in time: before the world was created, or during its first stages—anyway, long ago. But what gives the myth an operative value is that the specific pattern described is everlasting: it explains the present and the past as well as the future.[39]

Although the notion of human-plant kinship is set down in myths from many hundreds of years prior, it still has direct relevance to the custodians of these mythological traditions. In contemporary Indigenous cultures, the trans-specific relationships of care and nurture, set down in the creation accounts, are still very much alive. In his wide-ranging survey of kinship, Sahlins notes that

> [t]he relevant Māori category of common belonging, *tupuna*, normally translated as "ancestor" or "grandfather", is classificatory, denoting an ancient with some legendary significance for current life. Johansen (1954: 148) observes that in traditional sagas, *tupuna* may refer to flies, whales, birds, trees, the canoe which brought a tribe to New Zealand, and Captain Cook (for example). All such beings—including what we deem inanimate "things"—are subjects who share essential attributes of common descent, kinship, and personhood with Māori people.[40]

In Aotearoa, the whakapapa of the trees and other plants is made explicit in John White's account of Tāne bringing down the trees from heaven—the only known account of this story. The trees and other plants are the direct descendants of Tāne, the *atua* of the forest, and their *whakapapa* reads like that of a human family:

> Tāne took Mu-mu-whango (gentle noise of the air) to wife, and begat the *totara*-tree. He took Pu-whaka-hara (great origin), and begat the kahika (a creeper or vine), and ake-rau-tangi-(ake, tree of the "weeping leaf").[41]

This myth recounts a shared ancestry that is ever present, a fact that was remarked upon by an early ethnographer Elsdon Best:

> When the Māori entered the forest he felt that he was among his own kindred, for had not trees and man a common origin, both being off-spring of Tāne? Hence he was among his own folk as it were, and that forest possessed a tapu life principle even as man does. Thus, when the Māori wished to fell a tree wherefrom to fashion a canoe or house timbers, for two reasons he was compelled to perform a placatory rite ere he could slay one of the offspring of Tane. He saw in the majestic trees living creatures of an elder branch of the great family.[42]

Across the Tasman Sea, in the Northern Territory of Australia, Mak Mak women Nancy Daiyi and Kathy Deveraux explain their kinship relation-ships with trees:

> This tree here, we call "uncle" this tree.... Stringybark is for the women, and woolybutt is for the men. They call it "uncle." So, we're not just

related to *ngirwatt* [totems] for animals. We've got relationships to trees too. That's Mum's uncle, stringybark.[43]

And, in Yanyuwa country, still in the Northern Territory, Annie Isaac tells of her kinship relationship with the grey mangrove:

> That tree, the grey mangrove, is my most senior paternal ancestor and we people of the Wuyaliya clan name ourselves as these people who are kin to the grey mangrove.[44]

These are just a few examples of living, breathing kinship relationships that link botanical mythology with the plants growing in the earth. Whereas Europeans like myself have largely lost (or perhaps forgotten) our kinship connections with plants, thousands of such relationships remain in Indigenous cultures.

These origin stories, briefly explored in this chapter, are diverse and divergent. Yet, all share the idea that plants both share a genealogy with human beings and are made of the same substance. The origin myths also indicate that plants have their own history and ancestry. Plants have their own ways of living in this world, which are not necessarily dependent on human beings. Plants have their own purposes, which are not necessarily the same as human purposes. The myths also demonstrate that for many cultures, human beings are not situated above the plants in a hierarchy of importance akin to the *great chain of being*.[45] Humans and plants are kin, with the ensuing relationships of care and respect that this entails.

These myths challenge our view of plants, and of a wider nature, as just a set of resources for our exclusive use. They also challenge our dominant relationship with plants and with land as based on ownership of *property*. After reading the following excerpts we may begin to think of plants as our much-used relations who merit more respect than we are used to showing them. Like those involved in advocating for the rights of nature,[46] we may even begin to question the basis of our claims to *own* these nonhuman lives that make our own lives possible.

ROOTS

FROM THE FLESH OF YMIR

Then said Gangleri: "These are great tidings which I now, hear; that is a wondrous great piece of craftsmanship, and cunningly made. How was the earth contrived?" And Harr answered: "She is ring-shaped without, and round about her without lieth the deep sea; and along the strand of that sea they gave lands to the races of giants for habitation. But on the inner earth they made a citadel round about the world against the hostility of the giants, and. for their citadel they raised up the brows of Ymir the giant, and called that place Midgard. They took also his brain and cast it in the air, and made from it the clouds, as is here said:

> Of Ymir's flesh the earth was fashioned.
> And of his sweat the sea;
> Crags of his bones, trees of his hair.
> And of his skull the sky.
> Then of his brows the blithe gods made
> Midgard for sons of men;
> And of his brain the bitter-mooded
> Clouds were all created."

Then said Gangleri: "Much indeed they had accomplished then, methinks, when earth and heaven were made, and the sun and the constellations of heaven were fixed, and division was made of days; now whence come the men that people the world?" And Harr answered: "When the sons of Borr were walking along the sea-strand, they found two trees, and took up the trees and shaped men of them: the first gave them spirit and life; the second, wit and feeling; the third, form, speech, hearing, and sight. They gave them clothing and names: the male was called Askr, and the female Embla, and of them was mankind begotten, which received a dwelling-place under Midgard. Next they made for themselves in the middle of the world a city which is called Asgard; men call it Troy. There dwelt the gods and their kindred; and many tidings and tales of it have come to pass both on earth and aloft." (*Gylfaginning*)[48]

ALBERT9
DVRER
NORICVS
FACIEBAT
1504

And God said, "Let the earth sprout vegetation, plants yielding seed, and fruit trees bearing fruit in which is their seed, each according to its kind, on the earth." and it was so. The earth brought forth vegetation, plants yielding seed according to their own kinds, and trees bearing fruit in which is their seed, each according to its kind. And God saw that it was good. and there was evening and there was morning, the third day…

Then God said, "Let us make man in our image, after our likeness. And let them have dominion over the fish of the sea and over the birds of the heavens and over the livestock and over all the earth and over every creeping thing that creeps on the earth."

So God created man in his own image, in the image of God he created him; male and female he created them.

And God blessed them. And God said to them, "Be fruitful and multiply and fill the earth and subdue it, and have dominion over the fish of the sea and over the birds of the heavens and over every living thing that moves on the earth." And God said, "Behold, I have given you every plant yielding seed that is on the face of all the earth, and every tree with seed in its fruit. You shall have them for food. And to every beast of the earth and to every bird of the heavens and to everything that creeps on the earth, everything that has the breath of life, I have given every green plant for food." And it was so. (Genesis 1)[47]

SONG OF PURUṢA

A thousand heads hath Puruṣa, a thousand eyes, a thousand feet.
On every side pervading earth he fills a space ten fingers wide.
This Puruṣa is all that yet hath been and all that is to be;
The Lord of Immortality which waxes greater still by food.
So mighty is his greatness; yea, greater than this is Puruṣa.
All creatures are one-fourth of him, three-fourths eternal life in heaven.
With three-fourths Puruṣa went up: one-fourth of him again was here.
Thence he strode out to every side over what eats not and what eats.
From him Virāj was born; again Puruṣa from Virāj was born.
As soon as he was born he spread eastward and westward o'er the earth.
When Gods prepared the sacrifice with Puruṣa as their offering,
Its oil was spring, the holy gift was autumn; summer was the wood.
They balmed as victim on the grass Puruṣa born in earliest time.
With him the Deities and all Sādhyas and Ṛsis sacrificed.
From that great general sacrifice the dripping fat was gathered up.
He formed the creatures of-the air, and animals both wild and tame.
From that great general sacrifice Ṛcas and Sāma-hymns were born:
Therefrom were spells and charms produced; the Yajus had its birth from it.

From it were horses born, from it all cattle with two rows of teeth:
From it were generated kine, from it the goats and sheep were born.
When they divided Puruṣa how many portions did they make?
What do they call his mouth, his arms? What do they call his thighs and feet?

The Brahman was his mouth, of both his arms was the Rājanya made.
His thighs became the Vaiśya, from his feet the Śūdra was produced.
The Moon was gendered from his mind, and from his eye the Sun had birth;
Indra and Agni from his mouth were born, and Vayu from his breath.
Forth from his navel came mid-air the sky was fashioned from his head
Earth from his feet, and from his ear the regions. Thus they formed the worlds.
Seven fencing-sticks had he, thrice seven layers of fuel were prepared,
When the Gods, offering sacrifice, bound, as their victim, Puruṣa.
Gods, sacrificing, sacrificed the victim these were the earliest holy ordinances.
The Mighty Ones attained the height of heaven, there where the Sādhyas, Gods
    of old, are dwelling. (*Rig Veda*)[49]

On the nature of the five classes of animals it says in revelation, that, when the primeval ox passed away, there where the marrow came out grain grew up, of fifty and five species, and twelve species of medicinal plants grew; as it says, that out of the marrow is every separate creature, every single thing whose lodgment is in the marrow. From the horns arose peas, from the nose the leek, from the blood the grapevine from which they make wine—on this account wine abounds with blood—from the lungs the rue-like herbs, from the middle of the heart, thyme for keeping away stench, and every one of the others as revealed in the Avesta. (*Bundahisn*)[50]

J. Mulder Sculp

When Pangu, the firstborn, was declining toward death, he transformed his body. His *qi* became the wind and clouds; his voice became the thunder; his left eye became the sun; his right eye became the moon; his four limbs and five extremities became the Four Directions and Five Peaks; hid blood and fluid became the Yang-zi and Yellow Rivers; his sinews and arteries became the veins of the earth; his muscles and flesh became the soil of the fields; his hair and beard became the stars and constellations; his skin and body-hair became the grasses and trees; his teeth and bones became the metals and minerals; his semen and marrow became pearls and gems; his sweat and secretions became the rain and mire. And the various bugs on his body, moved by the wind, were transformed into the black-haired peasants. (*San Wu liji*, Xu Zheng)[51]

PELLERVOINEN SCATTERS SEEDS

Then did Väinämöinen, rising,
Set his feet upon the surface
Of a sea-encircled island,
In a region bare of forest.

There he dwelt, while years passed over,
And his dwelling he established
On the silent, voiceless island,
In a barren, treeless country.

Then he pondered and reflected,
In his mind he turned it over,
"Who shall sow this barren country,
Thickly scattering seeds around him?"

Pellervoinen, earth-begotten,
Sampsa, youth of smallest stature,
Came to sow the barren country,
Thickly scattering seeds around him.
Down he stooped the seeds to scatter,

On the land and in the marshes,
Both in flat and sandy regions,
And in hard and rocky places.
On the hills he sowed the pine-trees,
On the knolls he sowed the fir-trees,
And in sandy places heather;
Leafy saplings in the valleys.

In the dales he sowed the birch-trees,
In the loose earth sowed the alders,
Where the ground was damp the cherries,
Likewise in the marshes, sallows.

Rowan-trees in holy places,
Willows in the fenny regions,
Juniper in stony districts,
Oaks upon the banks of rivers.

Now the trees sprang up and flourished,
And the saplings sprouted bravely.
With their bloom the firs were loaded,
And the pines their boughs extended.
In the dales the birch was sprouting,
In the loose earth rose the alders,
Where the ground was damp the cherries,
Juniper in stony districts,
Loaded with its lovely berries;
And the cherries likewise fruited.

Väinämöinen, old and steadfast,
Came to view the work in progress,
Where the land was sown by Sampsa,
And where Pellervoinen laboured.
While he saw the trees had flourished,
And the saplings sprouted bravely,
Yet had Jumala's tree, the oak-tree,
Not struck down its root and sprouted.

Therefore to its fate he left it,
Left it to enjoy its freedom,
And he waited three nights longer,
And as many days he waited.
Then he went and gazed around him,
When the week was quite completed.
Yet had Jumala's tree, the oak-tree,
Not struck down its root and sprouted. (*The Kalevala*)[52]

Baccæ Iuniperi maioris.

IVNIPERVS
MINOR.

Weckholder.

du genefure

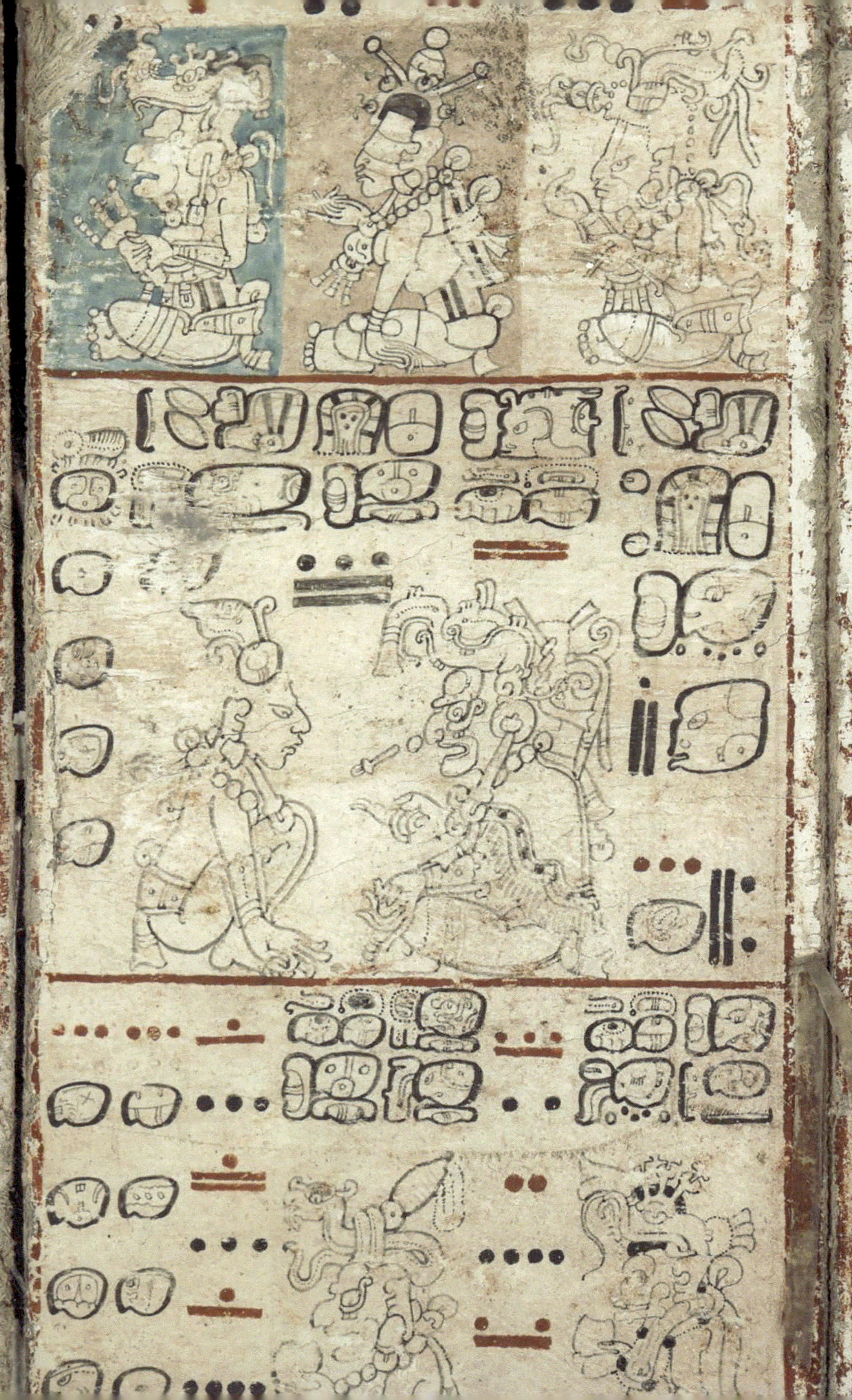

It was just before the sun, the moon, and the stars appeared over the creators Tepeu and Gucumatz. From Paxil, from Cayalá, came the yellow ears of corn and the white ears of corn. Four animals brought the yellow ears of corn and the white ears of corn, the *yac* (mountain cat), *utiú* (coyote), *quel* (small parrot), and *hoh* (crow). They told them that they should go to Paxil and showed them the road. The creators found the food, and this was what went into the flesh of created man. This corn was his blood and became part of his substance.

And in this way they were filled with joy, because they had found a beautiful land, full of pleasures, abundant in ears of yellow corn and ears of white corn, and abundant also in *pataxte* and *cacao*, and in innumerable *zapotes, anonas, jocotes, nantzes, matasanos,* and honey. There was an abundance of delicious food in the villages of Paxil and Cayalá. There were foods of every kind, small and large foods, small plants and large plants. The four animals showed them the road.

Then grinding the yellow corn and the white corn, Xmucané made nine drinks, and from this food came the strength and the flesh, and with it they created the muscles and the strength of man. This did the forefathers, Tepeu and Gucumatz. After that they began to talk about the creation of our first mother and father. They made their flesh from yellow corn and white corn; from corn meal dough they made the arms and the legs of man. Only dough of corn meal went into the flesh of our ancestors … (*Popol Vuh*)[53]

In the beginning two female human beings were born. These two children were born underground at a place called Shipapu. As they grew up, they began to be aware of each other. There was no light and they could only feel each other. Being in the dark they grew slowly.

After they had grown considerably, a Spirit whom they afterward called Tsichtinako spoke to them, and they found that it would give them nourishment. After they had grown large enough to think for themselves, they spoke to the Spirit when it had come to them one day and asked it to make itself known to them and to say whether it was male or female, but it replied only that it was not allowed to meet with them. They then asked why they were living in the dark without knowing each other by name, but the Spirit answered that they were *nuk'timi* (under the earth); but they were to be patient in waiting until everything was ready for them to go up into the light. So they waited a long time, and as they grew they learned their language from Tsichtinako.

When all was ready, they found a present from Tsichtinako, two baskets of seeds and little images of all the different animals (there were to be) in the world. The Spirit said they were sent by their father. They asked who was meant by their father, and Tsichtinako replied that his name was Ūch'tsiti and that he wished them to take their baskets out into the light, when the time came. Tsichtinako instructed them, "You will find the seeds of four kinds of pine trees, *lā'khok*, *gēi'etsu* (*dyai'its*), *wanūka*, and *lă'nye*, in your baskets. You are to plant these seeds and will use the trees to get up into the light." They could not see the things in their baskets but feeling each object in turn they asked, "Is this it?" until the seeds were found. They then planted the seeds as Tsichtinako instructed. All of the four seeds sprouted, but in the darkness the trees grew very slowly and the two sisters became very anxious to reach the light as they waited this long time. They slept for many years as they had no use for eyes. Each time they awoke they would feel the trees to see how they were growing. The tree *lanye* grew faster than the others and after a very long time pushed a hole through the earth for them and let in a very little light. The others stopped growing, at various heights, when this happened. (Acoma Origin Myth)[54]

TIGER SHARK BRINGS THE CYCADS

The Tiger Shark was in a country called Dumbarra in the east where he looked after the cycad palms that lived there. While he cared for the cycads, the other fish and sharks laughed at his large penis. As his retribution for the insult, the Tiger Shark gathered up all the cycads and took them from them.

He ripped up the cycads growing in the ground and even unearthed the cycad food being prepared in the ground. He put the cycad palms in a bundle on his back and silently swam across the neighboring country in search of country for the Rrumburriya people. After swimming a long time he at last found the country and placed a small bundle of cycad nuts on a reef which he named Wurlma.

Travelling south he arrived at a place he named as Nungkariwurra (an island). There he saw a rock wallaby who asked him what he was carrying. Tiger shark told her that it was the cycads which he had carried from the east, from Dumbarra. He asked the rock wallaby if he could place the bundle there as his shoulders had cramped from carrying the heavy bundle such a way. Angrily, the rock wallaby refused to take the cycad food. She threatened the Tiger Shark and shooed him away to take the food to the mainland. The Tiger Shark had no choice and wearily continued carrying the bundle south until he saw land, whereupon he threw the cycad nuts to the Wuyaliya and Rrumburriya country and travelled up the Wearyan River to Manankurra. (Yanyuwa Dreaming)[55]

PLANTING IN THE DREAMING

The first being of the creative era was a female, Warramurrungundji, who came out of the sea on the northern coast, in the Malay Bay. From there she moved through the Alligator Rivers region, forming the landscape, creating populations and planting foods for them. At the locality where she first entered the mainland, she dug two wells and planted *anbonde*, a banyan tree (*Ficus virens*). In the nearby freshwater swamp she scattered the seeds of *anguladji*, the spike rush (*Eleocharis dulcis*), whose rhizomes are a good food source for people and magpie geese. Further south, on the East Alligator River plains she again planted *anguladji* and also the seeds of *anrrol bunda*, the wild rice (*Oryza perennis*)." (Mayali Dreaming)[56]

Scale of one foot.

Scirpus tuberosus

It was Paia and his companions who separated Raki and Papa, and when they took him up also carried the trees, herbage, and edible roots with them, leaving Papa to lie naked. On looking down and seeing how bare Papa was, Tāne and Paia descended, and Tāne went out towards the sun (east-ward), to other settlements, to bring herbage and trees and other vegetation. He obtained some of each and every variety that grows, and from every district on the earth, and distributed them over every part of Papa, even to Ao-tea-roa (long light day), and Ta-ranga (repeating incantation), and Wai-roa-maire-he (long cadence of the evil song).

He classified the trees…

Tāne went far out, and brought the cod-fish (*hapuku*) from Te-ao-o-wai-raki-a-ira (the clear calm water of Ira; ira, spot on the skin, pimple, wart) as food to be in constant supply; and from the same place Raki and Taka-roa brought the baracouta (*mangaa*)—it came in summer and went back in winter. All fish of the sea came from the same place.

Tāne also obtained the *tio* (oyster), the *pipi* (cockle), the *paua* (*haliotis*), the *kakahi* (*unio*), the pupu (periwinkle), the karuru, the kareko (edible sea-weed that grows on the stones in water in the third Maori month), the *kapiti*, the *kauru* (*tii*-root).

When Tāne had done this, and clothed Papa, he disappeared by going up to heaven. Te-Rara-tau-karere-o-mati-te-raki is the name of the place whence Tāne brought trees, and took them to Huka-huka-te-raki (fringe of heaven), to Hu-matao (rather cold), to Tu-kou-a-hao-a-iki (nakedly standing, the gatherer-together and consumer), and to Ao-tea-mua (sacred light cloud).

When Tāne planted trees at Ao-tea-mua, he set the feet and legs in the earth—trees at first were like men—and retired a little distance to survey them; but they did not please him. He then planted the head downwards and the legs upwards, which he pronounced good: thus the hair of the head became the roots. Raki had little to do with them, though they were his children.

Te-ku-whaka-hara (the great coo of the bird) was the mother of the *totara* tree, and Te Kui-u-uku (old woman of the wiped breasts) of the *matai* tree, Ku-raki (coo of the bird to the north) of the *kahika* (koroi, or white pine), Huri-mai-te-ata (the dawn turning back) the *kahika-toa* (*manuka*, or tea-tree). (John White, *Ancient History of the Māori*)[57]

Dammara Australis.

When Iso-takeru no Kami descended from Heaven, he took down with him the seeds of trees in great quantity. However, he did not plant them in the land of Han [China], but brought them all back again, and finally sowed them every one throughout the Great Eight-island-country, beginning with Tsukushi. Thus green mountains were produced. For this reason Iso-takeru no Mikoto was styled Isaoshi no Kami. "He is the Great Deity who dwells in the Land of Kii."

In one writing it is stated:—"Sosa no wo no Mikoto said:—In the region of the Land of Han there is gold and silver. It will not be well if the country ruled by my son should not possess floating riches. So he plucked out his beard and scattered it. Thereupon Cryptomerias were produced. Moreover, he plucked out the hairs of his breast, which became Thuyas.' The hairs of his buttocks became Podocarpi. The hairs of his eye-brows became Camphor-trees. Having done so, he determined their uses. These two trees, viz. the Cryptomeria and the Camphor-tree, were to be made into floating riches; the Thuya was to be used as timber for building fair palaces; the Podocarpus was to form receptacles in which the visible race of man was to be laid in secluded burial-places. For their food he well sowed and made to grow all the eighty kinds of fruit.

Now the children of Sosa no wo no Mikoto were named Iso-takeru no Mikoto, with Oho-ya tsu hime, his younger sister, and next Tsuma'-tsu-hime no Mikoto. All these three Deities also dispersed well the seeds of trees, and forthwith crossed over to the Land of Kit [Korea]. (*Nihongi*)[58]

There is nothing that calls for notice
among these tribes individually, except
the worship of Nerthus or Mother Earth,
which is common to them all, the goddess,
according to their belief, mingling in the
affairs of men, and visiting her various
peoples in her chariot. On an island out
in the ocean there is an inviolate grove,
where, covered by a robe, is a sacred
car dedicated to her.

—Tacitus, *Germania*

2

GODS

THROUGHOUT HISTORY, a cast-iron way to label people as backward, if not a little bit crazy, has been to insinuate that they worship plants. When chronicling their empire, the Romans did their best to besmirch the European pagans by emphasizing a weird and wonderful attachment to sacred groves and arcane rituals. The classical Roman historian Claudian cast aspersions on the peoples of central Europe in his description of "the woods that old-established superstition has rendered awful, our axes fell the trees the barbarian once worshipped,"[1] and the great historian of the Roman Empire, Tacitus, described the Germanic peoples as naive worshippers of Mother Earth. Pliny's description of the Druids harvesting the sacred mistletoe also portrays pagan Europe in a similar vein (see chapter 4).

Such an approach was also taken in the Hebrew Bible. Old Testament texts refer to the mysterious goddess Asherah[2] and the votive objects set up to worship her. These objects are closely connected with living trees—described as being created through planting and frequently located "on every high hill and under every green tree"[3]—and the Bible contains frequent exhortations to cut down, destroy, and uproot them. The worship of Asherah's sacred trees is not only regarded as backward, but also as idolatrous. The book of Isaiah puts us on notice that

> [y]ou will be ashamed because of the sacred oaks in which you have delighted, you will be disgraced because of the gardens that you have chosen.[4]

To the Roman Imperialist and aspiring monotheists, there is clearly something unnerving about peoples who have relationships with plants that transcend *use*. Something so unnerving that the pioneering anthropologists E. B. Tylor and James Frazer felt the need to deem whole cultures to be primitive and backward on the basis that they worshipped sacred plants.[5] Is it not coincidental that both Tylor and Frazer, for whom the notion of sacred plants was largely symbolic—that is, plants acting as symbols for some higher, human truth—were Christians?

Sacred plants in Christianity are generally of this human-symbolic type.[6] An obvious example of a symbolic plant is the tree of the knowledge of good and evil found in Genesis. There is the vine that represents the relationship between the Hebrew God and his children, and also the passionflower (*Passiflora*), symbolizing the Passion of Christ. Some biblical plants are deemed sacred due to their association with Jesus, such as the palm tree via the travails of Palm Sunday or the Star of Bethlehem (*Ornithogalum umbellatum*) said to represent the star that indicated his birth.

This notion of sacred, as connected with religious practice or truth, and somehow *set apart* is a common one. Thinking of the *sacred* in binary opposition to ordinary life has a long history in the study of religion.[7] However, thinking of plant mythology in these terms perpetuates an anthropocentric view of plant life—the actual presence of plants is subsumed under human religious symbolism. They are contained tightly within our framework of *use*-based relationships, with the humans central and in control. Breaking out of this frame of reference is perhaps what was so unnerving about the idea of plant worship to scholars such as Tylor. Certainly, non–use-based relationships with plants appear primitive if you have a foundational philosophical idea that plants are purely objects/symbols for human consumption.

Yet, rather than portraying simple peoples naively getting on bended knee in the forest, botanical myths demonstrate a number of different types of relationships with sacred plants. It would pay to be mindful of these myths when thinking about how we might recast our own relationships with the plant kingdom. As for the representations of kinship in chapter 1, these depictions of alternative relationships are many and varied.

THE WORLD TREES

The *Upaniṣads* tell of the *asvattha* tree (also known as the *pipal* tree to Hindus and the Bodhi tree to Buddhists) (*Ficus religiosa*), a majestic, spreading tree, regarded as occupying the center of existence:

> There is that ancient tree whose roots grow upward and whose branches
> grow downward;—that indeed is called the Bright, that is called
> Brahman, that alone is called the Immortal. All worlds are contained in
> it, and no one goes beyond. This is that.[8]

Another world tree, the most famous tree of Norse mythology, Yggdrasil, first appears in the *Völuspá* as an "evergreen" ash tree:

> I know an ash standing
> Yggdrasil hight,
> a lofty tree, laved
> with limpid water:
> thence come the dews
> into the dales that fall;
> ever stands it green
> over Urd's fountain.[9]

In the later text, *Gylfaginning,* Yggdrasil's "world tree" status is developed further:

> The Ash is greatest of all trees and best: its limbs spread out over all the world and stand above heaven. Three roots of the tree uphold it and stand exceeding broad.[10]

Similarly, in the Karelian mythology of Finland, the *Kalevala* describes another northern European "world tree," the mighty Oak tree of Jumala (god of the sky), which connects the heavens and earth:

> Tender germ, of oak an acorn
> Whence the beauteous plant sprang upward,
> And the sapling grew and flourished,
> As from earth a strawberry rises,
> And it forked in both directions.
>
> Then the branches wide extended,
> And the leaves were thickly scattered,
> And the summit rose to heaven,
> And its leaves in air expanded.[11]

The seventeenth/eighteenth-century Mayan texts, the *Books of Chilam Balam*, contain a description of yet another world tree, *yaxche* (*Ceiba pentandra*). In the *Chilam Balam of Chumayel*, there is a description of creation—following the establishment of the Ceiba trees at all the cardinal points, "the green tree of abundance was set up in the centre <of the world> as a record of the destruction of the world."[12]

The historian of religion Mircea Eliade considered such trees as *yaxche* to be "world trees," symbols of "the living cosmos, endlessly renewing itself."[13] From Eliade's point of view, the *asvattha* tree is predominantly a symbol of

Brahman, the underlying nature of existence. Such a view has both commonalities with and differences from the Christian plant-symbolism discussed above. On one hand, the *asvattha* is a symbol for a religious concept, on the other, unlike in Christian symbolism, that concept extends beyond humanity into the fundamental nature of the universe.

A different take on world trees such as the *asvattha* has emerged from recent scholarship. In a fascinating work, the anthropologist David Haberman has written of the "worship" (*puja*) of trees, particularly the *asvattha*, in contemporary India. Unlike anthropologists before him, Haberman has positioned this "worship" in light of the construction of meaningful relationships with sentient plants that make our life on Earth possible.[14] Keeping this perspective, we could say that the *asvattha* is "sacred," not (or not only) because it's a symbol of Brahman, or because of its association with the enlightenment of the Buddha, but because of its life-giving role and its participation in relationships of nurture, love, and care.

According to Eliade's interpretation, in these excerpts, the roots of the broad ash tree, the branches of the oak, and the *yaxche*, symbolize the entire living world. In contrast, taking cues from Haberman's more recent work in India, there are others way of viewing the existence of these world trees. Rather than being stand-ins for the abstract notion of "the living cosmos, endlessly renewing itself," from a botanical perspective the above descriptions of the world trees actually contain a sense of a trees' embodied presence. In the above excerpts, the *asvattha*, Yggdrasil, the Oak tree of Jumala, and the *yaxche* are not just symbols for an abstract human truth. Within the descriptions of these all-encompassing trees, the sense of the trees' presence, their corporeal existence, is palpable and powerful. If these trees are symbols, they are symbols for the living botanical presence that makes life on Earth possible. In the case of *yaxche* it is a living presence which is intimately connected to the flourishing or destruction of the world. In all world trees it is this animate, life-giving presence that is sacred.

All the "world tree" passages above describe real trees, in rich, earthly detail. In the case of the Oak tree of Jumala, there is recognition that plants must die for other life to exist. As the tree shoots up and begins to block out the life beneath, the hero Väinämöinen must cut it down in order that life on earth continues to flourish. The sacredness of the tree is closely connected to nurture, and the sacrifice of botanical life for the sake of a wider, ecological flourishing.

In the case of Yggdrasil, the myth acknowledges that such a role involves sacrifice for the tree itself, because Yggdrasil is sentient and is capable of suffering—"Yggdrasil's ash hardship suffers greater than men know of."[15] This vulnerability is emphasized by Michael Bintley, whose survey of plant life in the *Poetic Edda* indicates that "[p]lants in eddic poetry experience suffering and loss in the same way as humans—in both physiological and psychological terms."[16] For Bintley:

> The experience of suffering is what really offers the best bridge between the sensory lives of plants and humans, given that plants have bodies, and often have limbs, which are just as prone to the violence of harsh weapons.[17]

Therefore, if Yggdrasil "operates as a *figura* for trees and plant life in more general terms, this represents an essential understanding within the Norse world-view that the life of trees is directly comparable with that of humans on certain basic levels."[18]

DIVINE ASSOCIATIONS

The recognition of world trees as *real trees,* exemplifying the living power of botanical life, puts a stop on the world trees' being backgrounded as simply symbols for human truths. This recognition also opens up the possibility of different types of relationships with these plants, relationships that transcend use and approach a connection based on gratitude and thanks. Scanning through those texts that make up the corpus of botanical mythology, the same can be said of other sacred plants, particularly those intimately associated with an aspect of the divine.[19]

In the Indian sacred text the *Padma Purana* there is a description of a divine association between the Indian sacred basil (*tulsi* or *tulasi*) and one of the three principal Hindu deities, Vishnu, the god of protection (and his incarnation Krishna). The god Shiva explains that the tulsi is sacred as it is the dwelling place of Krishna:

> The Tulsi plant is supreme among all the vegetations. She is very dear to lord Vishnu and fulfils all the desires of a man. Lord Krishna dwells near the place where there is a Tulsi plant.[20]

The most "worshipped" Indian tree of all, the *asvattha*, is sometimes considered to be the dwelling place (but more often the actual embodiment) of Vishnu (or sometimes Krishna). The text of the *Skanda Purana* relates that the "tree itself is a form of Vishnu" and this is still believed today—as David Haberman was told, "The tree is believed to be Vishnu himself."[21] The tree is worshipped at tree shrines as it is an embodiment of Vishnu, but this does not smother the corporeal presence of the plant itself. The worshippers of the *asvattha* tree all share the notion that not only is the plant an embodiment of divinity, but it is also a sentient, conscious being, worthy of respect in its own right.[22]

Many plants are considered sacred because they are actually made from a divine substance. In the *Padma Purana* there is a description of the creation of the *rudraksha* tree (*Elaeocarpus ganitrus*) whose seeds are traditionally used in the making of prayer beads in Hinduism and Buddhism. The lineage of the tree was created when a

> few drops of sweat, emanating from Shiva's body fell down on the earth.
> These sweat drops got transformed into a large Rudraksha-tree.[23]

The tree is not a symbol of the god Shiva. Created from the body of Shiva, the tree and the god share the same substance. The tree and Shiva share a mutuality of being that is the hallmark of kinship. Similarly, in another passage about *tulsi* (*tulasi*) from the *Devi Bhagavatam Purana*, the plant has its origins in the goddess Lakshmi, another metamorphosis from the hair of a divine being.[24]

> Your hairs will be turned into sacred trees and as they will be born of
> you, you will be known by the name of Tulasi. All the three worlds will
> perform their Pûjâs with the leaves and flowers of this Tulasi. Therefore,
> O Fair-faced One! This Tulasi will be reckoned as the chief amongst
> all flowers and leaves. In Heavens, earth, and the nether regions, and
> before Me, O Fair One, you will reign as the chief amongst trees and
> flowers.[25]

Interestingly, in the case of the sacred lotus (*Nelumbo nucifera*), the depiction of shared substance is reversed. In the story of the origins of the revered Tibetan Buddhist master Padmasambhava (also known to Tibetan Buddhists as Padmakara, who is said to have brought Buddhism to Tibet from India in the eighth century), it is the flower that gives birth to the human being. The story of the origins of Padmasambhava in the sacred lotus crosscuts the themes presented in this book—establishing a bond of kinship between sacred plants and human beings through an act of metamorphosis.

The idea of sentient plants as sharing substance with human beings or deities extends far beyond India. In ancient Egypt, the tree known as the sycamore (*Ficus sycomorus*) is often deemed sacred due to its connections with a number of Egyptian deities. In the *Payprus of Ani* (also known as the *Egyptian Book of the Dead*), the sycamore is considered to be the home of the goddess Nut, goddess of the sky. In the text, Ani, the Theban scribe for whom the papyrus was written proclaims, "Hail, thou sycamore tree of the goddess Nut! Grant thou to me of the water and the air which are in thee."[26] The text appears beside a vignette that shows the goddess Nut in a sycamore tree, pouring water into Ani's hands from a vessel. As for the *asvattha*, the sycamore is sacred as the *home* of Nut, but the earthly, sentient nature of the tree is not denied. Ani's proclamation of praise is directed at the sycamore tree, not at the goddess. The sacred tree is addressed as a being in its own right, a being that makes life on Earth possible. The sacredness of the sycamore is part of the tree's own nature, and is not simply due to its association with Nut. The connections between the life-giving properties and sacredness of the sycamore are also borne out in its relationship with the goddess Hathor, "lady of the sycamores" who, among other things, personifies motherhood and fertility.[27]

A similar connection between a deity and a sacred plant occurs in the mythology of the oaks of Dodona that are connected to Zeus, the ancient Greek god of the sky. Herodotus's account of the myth tells that

> two black doves had come flying from Thebes in Egypt, one to Libya and one to Dodona; this last settled on an oak tree, and uttered there human speech, declaring that there must be there a place of divination from Zeus; the people of Dodona understood that the message was divine, and therefore they established the oracular shrine.[28]

Once again, the human relationships here are with animate trees, not purely with embodied divinity. In communication with the humans around them, the tree itself has a presence, opening up the possibility of a relationship based on something other than *just* use. We'll explore these relationships in chapter 5.

USE OF THE SACRED

The status of plants as kin leads to respect and veneration, but it is important to be clear that what flows from this respect is care and consideration rather than naive supplication. For Gagudju elder Bill Neidjie, whose home country was in Kakadu National Park, in the Northern Territory of Australia, care for plants means:

Don't be rough!
If you too rough …
little bit of mistake.[29]

For me, being rough is running roughshod over plants—taking them for granted, ignoring their own lives, taking too much. Respect may be doing the opposite, but although they may be respected, many sacred plants are not excluded from the ranks of those plants that are acceptable to use. Some totemic trees are sacrosanct and should not be chopped down, but in many cultures, sacred plants are heavily used, forming part of everyday life and ritual practices.

In the Japanese epic the *Kojiki*, the sacred *saka-ki* trees have a divine origin and are kin to human beings, but their bark is also used for divination purposes.

> The eight hundred myriad deities assembled in a divine assembly in the bed of the tranquil river of heaven … and took cherrybark from the heavenly Mount Kagu.
>
> They performed divinations by pulling up from its roots a true saka-ki tree with five hundred branches, from the heavenly Mount Kagu. The deities put upon its upper branches a string of five hundred curved jewels which was eight feet long, and tied an eight feet long mirror to the middle branches. On the lower branches they hung white and blue peace offerings.[30]

The tobacco plant (*Nicotiana tabacum*) is sacred to many Indigenous North American peoples, while at the same time having important ceremonial uses. The tobacco plant is related to as a sentient being, the smoke of which is used to increase connectedness between the human beings that share the pipe.[31] Centuries ago, the *Natural History* introduced a similar account of the Druids who "hold nothing more sacred than mistletoe and a tree on which it is growing,"[32] but who also harvested mistletoe for use from the host oak tree.

> A priest arrayed in white vestments climbs the tree and with a golden sickle cuts down the mistletoe, which is caught in a white cloak.[33]

In a similar vein, cannabis is considered sacred in India, but also has well known medicinal and recreational uses. Indeed, one of the first known descriptions of the uses of the cannabis plant is from Herodotus, who describes the Scythians using the cannabis plant in a nice steam bath.

The Scythians then take the seed of this hemp and, creeping under the rugs, they throw it on the red-hot stones; and, being so thrown, it smoulders and sends forth so much steam that no Greek vapour-bath could surpass it. The Scythians howl in joy for the vapour-bath. This serves them instead of bathing, for scarce ever do they wash their bodies with water.[34]

What's more, perhaps the most well-known of all sacred plants, the *soma* found in the Indian Vedic literature, is regularly incorporated into religious rituals, which involve the use of the sacred plant.

We have drunk Soma and become immortal; we have attained the light, the Gods discovered.... As a wise Friend to friend: do thou, wide-ruler, O Soma, lengthen out our days for living.[35]

Although the *soma* plant is very useful for enabling man to "discover the gods," it is also addressed directly and respectfully as a sentient being; both a "friend" and as the "ruler" of the plant kingdom. The relationship with soma is not restricted to its ceremonial use and the *soma* plant is not reduced to an instrument of human purpose.

GODS

IN THE GARDEN OF EDEN

When no bush of the field was yet in the land and no small plant of the field had yet sprung up—for the Lord God had not caused it to rain on the land, and there was no man to work the ground, and a mist was going up from the land and was watering the whole face of the ground—then the Lord God formed the man of dust from the ground and breathed into his nostrils the breath of life, and the man became a living creature. And the Lord God planted a garden in Eden, in the east, and there he put the man whom he had formed. And out of the ground the Lord God made to spring up every tree that is pleasant to the sight and good for food. The tree of life was in the midst of the garden, and the tree of the knowledge of good and evil. (Genesis 2)[36]

J. Mulder Sculp.

The hero sauntered along the banks of the river. Night slowly descended. The flowers wearily closed their petals; a sweet fragrance rose from the fields and gardens; the birds timidly rehearsed their evensongs.

It was then the hero walked toward the tree of knowledge.

The road was sprinkled with gold-dust; rare palms, covered with precious stones, lined the way. He skirted the edge of a pool whose blessed waters exhaled an intoxicating perfume. White, yellow, blue and red lotuses spread their massive petals over the surface, and the air rang with the clear songs of the swans. Near the pool, under the palms, Apsarases were dancing, while in the sky the Gods were admiring the hero.

He approached the tree. On the side of the road, he saw Svastika, the reaper.

"They are tender, these grasses you are mowing, Svastika. Give me some grass; I want to cover the seat I shall occupy when I attain supreme knowledge. They are green, these grasses you are mowing, Svastika. Give me some grass, and you will know the law some day, for I shall teach it to you, and you may teach it to others."

The reaper gave the Saint eight handfuls of grass. There stood the tree of knowledge. The hero went to the east of it and bowed seven times. He threw the handfuls of grass on the ground, and, suddenly, a great seat appeared. The soft grass covered it like a carpet. The hero sat down, his head and shoulders erect, his face turned to the east. Then he said in a solemn voice:

"Even if my skin should parch, even if my hand should wither, even if my bones should crumble into dust, until I have attained supreme knowledge I shall not move from this seat."

And he crossed his legs. (*The Life of the Buddha*)[37]

Then said Gangleri: "Where is the chief abode or holy place of the gods?" Hárr answered: "That is at the Ash of Yggdrasill; there the gods must give judgment every day." Then Gangleri asked: "What is to be said concerning that place?" Then said Jafnharr: "The Ash is greatest of all trees and best: its limbs spread out over all the world and stand above heaven. Three roots of the tree uphold it and stand exceeding broad: one is among the Æsir; another among the Rime-Giants, in the place where aforetime was the Yawning Void; the third stands over Niflheim, and under that root is Hvergelmir, and Nídhöggr gnaws the root from below. But under that root which turns toward the Rime-Giants is Mímir's Well, wherein wisdom and understanding are stored; and he is called Mímir, who keeps the well. He is full of ancient lore, since he drinks of the well from the Gjallar-Horn. Thither came Allfather and craved one drink of the well; but he got it not until he had laid his eye in pledge. So says Völuspá:

> All know I, Odin, where the eye thou hiddest,
> In the wide-renowned well of Mímir;
> Mímir drinks mead every morning
> From Valfather's wage. Wit ye yet, or what?

The third root of the Ash stands in heaven; and under that root is the well which is very holy, that is called the Well of Urdd; there the gods hold their tribunal. Each day the Æsir ride thither up over Bifrost, which is also called the Æsir's Bridge. These are the names of the Æsir's steeds: Sleipnir is best, which Odin has; he has eight feet. The second is Gladr, the third Gyllir, the fourth Glenr, the fifth Skeidbrimir, the sixth Silfrintoppr, the seventh Sinir, the eighth Gísl, the ninth Falhófnir, the tenth Gulltoppr, the eleventh Léttfeti. Baldr's horse was burnt with him; and Thor walks to the judgment, and wades those rivers which are called thus:

> Körmt and Örmt and the Kerlaugs twain,
> Them shall Thor wade
> Every day when he goes to doom
> At Ash Yggdrasill;
> For the Æsir's Bridge burns all with flame,
> And the holy waters howl."…

Vedr falner
Palrm mylla augna
Orn hedur i fiun oskil
Duneyz
Dualin
Dain
Dyrafzor
Rata
tøskur
ber op
undar
ord my
llu akr
og eyst
hoggs
Nydhøgg
nagar
44

Then said Gangleri: "What more mighty wonders are to be told of the Ash?" Hárr replied: "Much is to be told of it. An eagle sits in the limbs of the Ash, and he has understanding of many a thing; and between his eyes sits the hawk that is called Vedrfölnir. The squirrel called Ratatöskr runs up and down the length of the Ash, bearing envious words between the eagle and Nídhöggr; and four harts run in the limbs of the Ash and bite the leaves. They are called thus: Dáinn, Dvalinn, Duneyrr, Durathrór. Moreover, so many serpents are in Hvergelmir with Nídhöggr, that no tongue can tell them, as is here said:

> Ash Yggdrasill suffers anguish,
> More than men know of:
> The stag bites above; on the side it rotteth,
> And Nídhöggrr gnaws from below.

And it is further said:

> More serpents lie under Yggdrasill's stock
> Than every unwise ape can think:
> Góinn and Móinn (they're Grafvitnir's sons),
> Grábakr and Grafvölludr;
> Ófnir and Sváfnir I think shall aye
> Tear the trunk's twigs.

It is further said that these Norns who dwell by the Well of Urdr take water of the well every day, and with it that clay which lies about the well, and sprinkle it over the Ash, to the end that its limbs shall not wither nor rot; for that water is so holy that all things which come there into the well become as white as the film which lies within the egg-shell,—as is here said:

> I know an Ash standing called Yggdrasill,
> A high tree sprinkled with snow-white clay;
> Thence come the dews in the dale that fall—
> It stands ever green above Urdr's Well.

> That dew which falls from it onto the earth is called by men honey-dew, and thereon are bees nourished. Two fowls are fed in Urdr's Well: they are called Swans, and from those fowls has come the race of birds which is so called." (*Gylfaginning*)[38]

OAK TREE OF JUMALA

Väinämöinen, old and steadfast,
Came to view the work in progress,
Where the land was sown by Sampsa,
And where Pellervoinen laboured.
While he saw the trees had flourished,
And the saplings sprouted bravely,
Yet had Jumala's tree, the oak-tree,
Not struck down its root and sprouted.

Therefore to its fate he left it,
Left it to enjoy its freedom,
And he waited three nights longer,
And as many days he waited.
Then he went and gazed around him,
When the week was quite completed.
Yet had Jumala's tree, the oak-tree,
Not struck down its root and sprouted.

Then he saw four lovely maidens;
Five, like brides, from water rising;

And they mowed the grassy meadow,
Down they cut the dewy herbage,
On the cloud-encompassed headland,
On the peaceful island's summit,
What they mowed, they raked together,
And in heaps the hay collected.

From the ocean rose up Tursas,
From the waves arose the hero,
And the heaps of hay he kindled,
And the flames arose in fury.

All was soon consumed to ashes,
Till the sparks were quite extinguished.
Then among the heaps of ashes,
In the dryness of the ashes,
There a tender germ he planted,
Tender germ, of oak an acorn
Whence the beauteous plant sprang upward,
And the sapling grew and flourished,
As from earth a strawberry rises,
And it forked in both directions.

Then the branches wide extended,
And the leaves were thickly scattered,
And the summit rose to heaven,
And its leaves in air expanded.

In their course the clouds it hindered,
And the driving clouds impeded,
And it hid the shining sunlight,
And the gleaming of the moonlight. (*The Kalevala*)[39]

<This is> the history of the world in those times, because it has been written down, because the time has not yet ended for making these books, these many explanations, so that Maya men may be asked if they know how they were born here in this country, when the land was founded.

It was <Katun> Ahau when the Ah Mucenca came forth to blindfold the faces of the Oxlahun-ti-ku; but they did not know his name, except for his older sister and his sons. They said his face had not yet been shown to them also. This was after the creation of the world had been completed, but they did not know it was about to occur. Then Oxlahun-ti-ku was seized by Bolon-ti-ku. Then it was that fire descended, then the rope descended, then rocks and trees descended. Then came the beating of <things> with wood and stone. Then Oxlahun-ti-ku was seized, his head was wounded, his face was buffeted, he was spit upon, and he was <thrown> on his back as well. After that he was despoiled of his insignia and his smut. Then shoots of the *yaxum* tree were taken. Also Lima beans were taken with crumbled tubercles, hearts of small squash-seeds, large squash-seeds and beans, all crushed. He wrapped up the seeds <composing> this first Bolon *¢acab*, and went to the thirteenth heaven. Then a mass of maize-dough with the tips of corn-cobs remained here on earth. Then its heart departed because of Oxlahun-ti-ku, but they did not know the heart of the tubercle was gone. After that the fatherless ones, the miserable ones, and those without husbands were all pierced through; they were alive though they had no hearts. Then they were buried in the sands, in the sea.

There would be a sudden rush of water when the theft of the insignia <of Oxlahun-ti-ku> occurred. Then the sky would fall, it would fall down upon the earth, when the four gods, the four Bacabs, were set up, who brought about the destruction of the world. Then, after the destruction of the world was completed, they placed a tree to set up in its order the yellow cock oriole. Then the white tree of abundance was set up. A pillar of the sky was set up, a sign of the destruction of the world; that was the white tree of abundance in the north.

Then the black tree of abundance was set up in the west for the black-breasted *pi¢oy* to sit upon. Then the yellow tree of abundance was set up in the south, as a symbol of the destruction of the world, for the yellow-breasted

Pub by S. Curtis, Glazenwood Essex Nov.r 1 1834.

picoy to sit upon, for the yellow cock *oriole* to sit upon, the yellow timid *mut*. Then the green tree of abundance [yaxche] was set up in the center of the world as a record of the destruction of the world. (*Chilam Balam of Chumayel*)[40]

During Satya-yuga, there lived a mighty demon named Tripurasur. He had con-quered the deities and was capable of moving in the space. The deities sought help of Lord Shiva, who killed Tripurasur by the sight of his third-eye. In the process, few drops of sweat, emanating from Shiva's body fell down on the earth. These sweat drops got transformed into a large Rudraksha-tree…

Once Kartikeya asked Lord Shiva about the holiest tree, which was capable of giving salvation. Lord shiva replied—The Tulsi plant is supreme among all the vegetations. She is very dear to lord Vishnu and fulfils all the desires of a man. Lord Krishna dwells near the place where there is a Tulsi plant. Spirits and ghosts never dare to venture near the Tulsi plant. If a man attaches a Tulsi leaf to his Shikha at the time of his death, he is liberated from all his sins. One who worships Lord Vishnu by offering Tulsi-leaves attains salvation. (*Padma Purana*)[41]

Vishnu & his favorite wife
Lakshmi

In the land of Uddiyana situated to the west of Bodhgaya there was an island in a huge lake on which appeared a multicolored lotus flower through the blessings of the Buddhas. Buddha Amitabha sent from his heart center a golden vajra marked with the letter HRIH into the bud of this lotus flower which miraculously turned into a small child eight years of age holding a vajra and a lotus and adorned with the major and minor marks. The child remained there teaching the profound Dharma to the devas and dakinis on the island.

At that time Indrabodhi, who was the king of the country, had no sons. He had already emptied out his treasury by making offerings to the Three Jewels and giving alms to the poor. As a last resort, in order to find a wish fulfilling jewel he embarked on a journey with his minister Krishnadhara on the great lake. On their return first Krishnadhara and later King Indrabodhi met the miraculous child. The king regarded him as an answer to his prayers for a son and brought him to the palace where he was given the name Padmakara, the Lotus Born. Padmakara was then asked to sit on a throne made of precious gems and given lavish offerings by all the people.

The prince grew up, bringing countless beings to maturation through his youthful sports and games.... In an act of play, he then pretended that his trident slipped out of his hand; it fell and killed the son of one of the ministers. He was then sentenced to be expelled to a charnel ground. He remained in Cool Grove, Joyful Forest and Sosaling, engaging in the conduct of yogic disciplines. During this time he received empowerment and blessings from the two dakinis Tamer of Mara and Sustainer of Bliss. When bringing all the dakinis of the charnel grounds under his command, he was known as Shantarakshita.

Padmakara returned to Uddiyana, to the island in Lake Danakosha where he practiced Secret Mantra and the symbolic language of the dakinis through which he brought the dakinis on the island under his command. He then practiced in the Rugged Forest and was blessed with a vision of Vajra Yogini. He bound under oath all the nagas of the lakes as well as the planetary spirits and was invested with supernatural powers by all the dakas and dakinis. Thus he became renowned as Dorje Drakpo Tsal, Wrathful Vajra Power. (Dakini Teachings)[42]

The Sacred Egyptian Bean

London, Published Dec.r 1. 1804, by D.r Thornton.

Eychbaum.
CXXIX.
Quercus Robur L.

They used to say, my friend, that the words of the oak in the holy place of Zeus at Dodona were the first prophetic utterances. The people of that time, not being so wise as you young folks, were content in their simplicity to hear an oak or a rock, provided only it spoke the truth; but to you, perhaps, it makes a difference who the speaker is and where he comes from, for you do not consider only whether his words are true or not. (Plato, *Phaedrus*)[43]

But whence each of the gods came into being, or whether they had all for ever existed, and what outward forms they had, the Greeks knew not till (so to say) a very little while ago; for I suppose that the time of Hesiod and Homer was not more than four hundred years before my own; and these are they who taught the Greeks of the descent of the gods, and gave to all their several names, and honours, and arts, and declared their outward forms. But those poets who are said to be older than Hesiod and Homer were, to my thinking, of later birth. The earlier part of all this is what the priestesses of Dodona tell; the later, that which concerns Hesiod and Homer, is what I myself say.

But as concerning the oracles in Hellas, and that one which is in Libya, this is the account given by the Egyptians. The priests of Zeus of Thebes told me that two priestesses had been carried away from Thebes by Phoenicians; one of them (so, they said, they had learnt) was taken away and sold in Libya, and the other in Hellas; these women, they said, were the first founders of places of divination in the countries aforesaid. When I asked them how it was that they could speak with so certain knowledge, they said in reply that their people had sought diligently for these women, and had never been able to find them, but had learnt later the tale which was now told to me.

That, then, I heard from the Theban priests; and what follows, is told by the prophetesses of Dodona: to wit, that two black doves had come flying from Thebes in Egypt, one to Libya and one to Dodona; this last settled on an oak tree, and uttered there human speech, declaring that there must be there a place of divination from Zeus; the people of Dodona understood that the message was divine, and therefore they established the oracular shrine. The dove which came to Libya bade the Libyans (so they say) to make an oracle of Ammon; this also is sacred to Zeus. Such was the tale told by the Dodonaean priestesses, of whom the eldest was Promeneia and the next in age Timarete, and the youngest Nicandra; and the rest of the servants of the temple at Dodona likewise held it true. (Herodotus, *Histories*)[44]

The eight hundred myriad deities assembled in a divine assembly in the bed of the tranquil river of heaven ... and took cherrybark from the heavenly Mount Kagu.

They performed divinations by pulling up from its roots a true saka-ki tree with five hundred branches, from the heavenly Mount Kagu. The deities put upon its upper branches a string of five hundred curved jewels which was eight feet long, and tied an eight feet long mirror to the middle branches. On the lower branches they hung white and blue peace offerings. (The *Kojiki*)[46]

Mistletoe berries can be used for making bird-lime, if gathered at harvest time while unripe; for if the rainy season has begun, although they get bigger in size they lose in viscosity. They are then dried and when quite dry pounded and stored in water, and in about twelve days they turn rotten—and this is the sole case of a thing that becomes attractive by rotting. Then after having been again pounded up they are put in running water and there lose their skins and become viscous in their inner flesh. This substance after being kneaded with oil is bird-lime, used for entangling birds' wings by contact with it when one wants to snare them.

While on this subject we also must not omit the respect shown to this plant by the Gallic provinces. The Druids—that is what they call their magicians— hold nothing more sacred than mistletoe and a tree on which it is growing, provided it is a hard-oak. Groves of hard-oaks are chosen even for their own sake, and the magicians perform no rites without using the foliage of those trees, so that it may be supposed that it is from this custom that they get their name of Druids, from the Greek word meaning "oak"; but further, anything growing on oak-trees they think to have been sent down from heaven, and to be a sign that the particular tree has been chosen by God himself. Mistletoe is, however, rather seldom found on a hard-oak, and when it is discovered it is gathered with great ceremony, and particularly on the sixth day of the moon (which for these tribes constitutes the beginning of the months and the years) and after every thirty years of a new generation, because it is then rising in strength and not one half of its full size. Hailing the moon in a native word that means "healing all things," they prepare a ritual sacrifice and banquet beneath a tree and bring up two white bulls, whose horns are bound for the first time on this occasion. A priest arrayed in white vestments climbs the tree and with a golden sickle cuts down the mistletoe, which is caught in a white cloak. Then finally they kill the victims, praying to God to render his gift propitious to those on whom he has bestowed it. They believe that mistletoe given in drink will impart fertility to any animal that is barren. and that it is an antidote for all poisons, So powerful is the superstition in regard to trifling matters that frequently prevails among the races of mankind. (Pliny, *Natural History*)[45]

We have drunk Soma and become immortal; we have attained the light, the Gods discovered.

Now what may foeman's malice do to harm us? What, O Immortal, mortal man's deception?

Absorbed into the heart, be sweet, O Indu, as a kind father to his son, O Soma,

As a wise Friend to friend: do thou, wide-ruler, O Soma, lengthen out our days for living.

These glorious drops that give me freedom have I drunk. Closely they knit my joints as straps secure a car.

Let them protect my foot from slipping on the way: yea, let the drops I drink preserve me from disease.

Make me shine bright like fire produced by friction: give us a clearer sight and make us better.

For in carouse I think of thee, O Soma, Shall I, as a rich man, attain to comfort?

May we enjoy with an enlivened spirt the juice thou givest like ancestral riches.

O Soma, King, prolong thou our existence as Sūrya makes the shining days grow longer.

King Soma, favour us and make us prosper: we are thy devotees; of this be mindful.

Spirit and power are fresh in us, O Indu give us not up unto our foeman's pleasure.

For thou hast settled in each joint, O Soma, aim of men's eyes and guardian of our bodies.

When we offend against thine holy statutes, as a kind Friend, God, best of all, be gracious.

May I be with the Friend whose heart is tender, who, Lord of Bays! when quaffed will never harm me—

This Soma now deposited within me. For this, I pray for longer life to Indra.

Our maladies have lost their strength and vanished: they feared, and passed away into the darkness.

Soma hath risen in us, exceeding mighty, and we are come where men prolong existence.

Fathers, that Indu which our hearts have drunken, Immortal in himself, hath entered mortals.

So let us serve this Soma with oblation, and rest securely in his grace and favour.

Associate with the Fathers thou, O Soma, hast spread thyself abroad through earth and heaven.

So with oblation let us serve thee, Indu, and so let us become the lords of riches,

Give us your blessing, O ye Gods' preservers. Never may sleep or idle talk control us.

But evermore may we, as friends of Soma, speak to the synod with brave sons around us.

On all sides, Soma, thou art our life-giver: aim of all eyes, light-finder, come within us.

Indu, of one accord with thy protections both from behind and from before preserve us. (*Rig Veda*)[47]

The poets also adorn
the falsehoods of error
by elegance of words,
and by sweetness of speech
persuade that mortals have been
made immortal; what's more,
they say that men are changed
into stars, and trees, and animals,
and flowers, and birds, and fountains,
and rivers. And but that it might
seem to be a waste of words, I could
even enumerate almost all the stars,
and trees, and fountains,and rivers, which
they assert to have been made of men; yet,
by way of example, I shall mention at least
one of each class. They say that … Hyacinthus,
beloved of Apollo,was turned into a flower.…
And they assert that almost all the stars,
trees, fountains, and rivers, flowers,
animals, and birds, were at
one time human beings.
—Clement of Alexandria,
Recognitions

3

METAMORPHOSIS

THE ANCIENT POETS came under heavy fire from the early Christian theologians for their heathen assertions that the human, animal, and vegetal realms interpenetrate. Faced with tales of transformations between man and beast, the founding church father, St. Augustine, responded with accusations of religious heresy, emphatic that "all those tales are lies."[1] Similarly, Clement of Alexandria, another great Christian theologian could barely disguise his disdain in the face of a poetic worldview in which "stars, trees, fountains and rivers … were at one time human beings."[2]

I don't feel a pressing urge to champion Christianity over pre-Christian paganisms, so we need not follow these early theological depictions of literal, individual transformations from human into animal and plant form. To my mind, Ovid puts forward a much more insightful understanding through the voice of the sage Pythagoras. In his definitive and hugely influential work, *Metamorphoses*, Ovid explains that the metamorphoses are not a theological or existential threat, nor an affront to rationality. They are simply a description of the natural and ever-occurring process of death and rebirth—an earthly life in which "Nothing persists without changing its outward appearance."[3] Mythical and poetic traditions from across the world have used stories of metamorphosis and change to highlight the natural, and ever-occurring transformation of form and energy between microbes, plants, animals, and humans. The metamorphosis is far from fantastical—it is the basis of the life cycle on planet Earth, a life cycle that has no end or beginning, but which does contain two fundamental markers—birth and death.

The mythological transformations that involve plants are focused on these two different aspects of earth-measured time. Sometimes they focus on birth, describing the origins of humankind, born directly out of the plant kingdom. More often, they appear to privilege death and the metamorphosis of human beings back into the Earth, specifically into botanical form. Although from a human centered perspective this often appears to be an abrupt ending, if we heed Pythagoras's words that "[e]verything changes and nothing can die,"[4] this is not what the philosopher Steven Luper calls "annihilation,"[5] but simply matter taking on a new shape, and/or souls finding new and receptive bodies. The end of this human existence is always the start of a different form of life.[6] In the wellspring of botanical mythology, this end of human life often leads to the very beginnings of plant lineages—of life forms, varieties, and species.

MORTAL MEN: SPRUNG FROM ASH TREES

In those botanical myths that privilege the idea of birth, we find stories that relate the origins of humankind itself. Stories excerpted in this volume describe a human species that was fashioned like a sculpture directly from the plant kingdom. In Northern Europe, the Old Norse poem *Völuspá* tells us that the first humans, Askr and Embla, were made from wood, crafted by the shaping of trees into human form. The gods Odin, Hœnir, and Lóðurr found the Ash and Elm and, after sculpting them into human shape, gave them spirit and blood.[7]

This direct crafting of the human species from trees is also found in the Quiche Mayan sacred text the *Popol Vuh*. Following the creation of the Earth and the coming forth of the mountains and trees, the forest animals, such as the birds, pumas, jaguars, and snakes, were created in order to guard the forest.

> Shall there be only silence and calm under the trees, under the vines? It
> is well that hereafter there be someone to guard them.[8]

Under the canopy, the animals scream and cackle and hiss, but when they are unable to the form words of praise that the Creators desire, the Creators decide to make human beings to fulfill this task. In their first attempt, they use earth as the building material, but it is too soft and just melts away. The Creators then turn to the soothsayers and the "Grandmother of the Dawn," who, in the midst of divination, recommends them to use wood—"Your figures of wood shall come out well; they shall speak and talk on earth."[9] The Creators sculpt the first humans out of wood, and, "They looked like men, talked like men, and populated the surface of the earth."[10] The text goes on to recount that the first humans of the present area were made from maize, with blood made from water. Other Mayan accounts relate the creation of the first humans from fodder, or thin reeds.[11]

Myths also tell us that human beings can be crafted from other plant lives. In the Welsh medieval epic the *Four Branches of the Mabinogi*, the woman Blodeuwedd is brought into being from a mixture of different blossoms, including broom and meadowsweet. On the surface, such direct formation of human flesh from plant matter seems remote and fantastical. Yet, looking deeper, it could be argued that these myths are relating events that occur every day. Whenever we eat a vegetable or a piece of fruit we start the process of transforming plant matter into the human cells that make up our own bodies. In the mythology recounted, this wondrous transformation is simply being acknowledged and championed. The myths of metamorphosis not only relate that plants are the foundation of our human forms, but also that plants are our progenitors. That is, like the stories in chapter 1, they recognize that the plant species which we live with and among are our ancestors and relatives.

In his *Works and Days*, the Ancient Greek poet Hesiod wrote of "mortal men, a brazen race, sprung from ash trees,"[12] echoing the Norse myth of the creation of mankind from the ash. Hesiod uses the word *meliai* to describe the ash trees from which men were formed—a term used to also describe the tree nymphs in ancient Greek literature (which are explored in chapter 5). Hesiod also provides the first explicit mention of these tree nymphs:

> All the bloody drops that gushed forth, Gaia received, and as the year moved round she bore the strong Erinyes and the great giants gleaming in their armour, with long spears in their hands, and the nymphs whom they call Meliai upon the boundless Earth.[13]

As the offspring of Gaia, according to Hesiod, the *meliai* nymphs (which themselves share a mutuality of being with the trees) and human beings are earth-born kin.

This notion of sylvan kinship is spread far and wide. In the *Aeneid*, Virgil describes humankind springing forth from the oak tree in the manner of a birth, while in the Americas, recorded tales of the Incas describe the first humans being created out of clay and then being born out of various orifices of the earth, including the trunks of trees.

> Thus they say that some came out of caves, others issued from hills, others from fountains, others from the trunks of trees.[14]

The idea that plants are the ancestors of humans may seem far-fetched if, like Clement of Alexandria, we consider the myths to be speaking of a literal, corporeal birth. However, if we keep in mind the common ancestors of the opening two chapters, the idea of plants being our direct relatives is not so fanciful. With our Darwinian understanding of the evolutionary process, it should also be largely self-evident that plants and humans are part of the same family tree. Indeed, advances in molecular phylogenetics indicate that plants and human beings are much closer kin than would at first appear. If the tree of life were a hand, the rice plant *Oryza sativa* and the human *Homo sapiens* would both occupy the end of a single fingernail.[15]

In these myths of metamorphosis, kinship relationships are not something that exist only in the abstract; they are real and corporeal connections, which take the familiar forms of human relationships.

In the Greek myth of the birth of the Adonis, the famously beautiful god is brought into the world from his mother Myrrha, the myrrh tree. Many artistic representations of the story show Adonis having emerged from the folds of bark, an infant in human form, with a mother who feels the pain and suffering of childbirth, but who is also a tree. In contrast to Robert Segal who regards Adonis's story in terms of child-parent archetypes,[16] in my interpretation, through this transformation of matter from plant to human, and the interpenetration of human and plant at birth, human beings and plants are displayed as having a deep and substantial kinship connection that transcends their temporary forms.

FINE BARK OVER SMOOTH SKIN

The same connections are also expressed in stories that relate the end of the human phase of the cycle, the transformation of human beings back into plants. In fact, the metamorphosis of humans into plants is one of the most common themes involving plants in Greek mythology. These metamorphoses are often connected to the nymphs, first mentioned by Hesiod.

The most celebrated and well-known example is Daphne's transformation into a laurel tree to escape the pursuing Apollo. When Daphne, having fled the amorous god, begins to tire, she issues a plea for help to her father (the river God Peneus), begging him to rid her of her beauty. As Ovid tells:

> Her prayer was scarcely finished when she feels
> a torpor take possession of her limbs—
> her supple trunk is girdled with a thin
> layer of fine bark over her smooth skin:
> her hair turns into foliage, her arms
> grow into branches, sluggish roots adhere
> to feet that were so recently so swift,
> her head becomes the summit of a tree:
> all that remains of her is a warm glow.[17]

According to myths, many plants once had this initial human form. Upon death, these humans transformed into some of our most familiar and common garden species (anemone, cypress, crocus, elm, heliotrope, hyacinth, iris, mint, and rose) and traces of this initial humanity can still be seen in their names.

The brightly colored petals of the hyacinth were once the beautiful youth Hyacinthos, another lover of Apollo, who was killed by a heavy discus flung by the amorous god.[18] At death, the blood that fell from Hyacinthos's head was transformed into the hyacinth and the petals of the flower bear the grief of the distraught Apollo.[19] The garden mint also has its origins in death, arising from the demise of the nymph Minthe who was trampled underfoot by Persephone, queen of the underworld and goddess of vegetation.[20] Numerous other myths describe the origin of a plant species following the death of a human being, including that of the origin of the anemone following the violent death of Adonis.[21]

The poplar tree is also a plant seemingly replete with human death and grief. The ancient Greek historian Diodorus Siculus is one of several sources (along with Ovid) of the myth of the Heliades, the sisters of the god Phaethon who are consumed by their grief on his death and are transformed into poplar trees. The tears of the poplar remain a reminder of their grief to this day,[22] a symbolism that is counterbalanced by the poplars' personhood. This link between familial death and the poplar tree also occurs in the myths of native North Americans. The Koyukon of Alaska tell of a woman who was transformed into a poplar tree on cutting her throat after the death of her husband the Raven.[23]

Such transformations from human into plant do not only occur in cases of spectacular deaths. In the Greek myth of Baucis and Philemon, the transformation is a divine reward. In Ovid's version of the myth, this old married couple were one day visited by the gods Zeus and Hermes. Although they lived in poverty, the couple showed great hospitality and provided the gods with a simple and honest meal. Having served in this way, they were granted exemption from the punishments being meted out on the surrounding area, and were rewarded for lives of honesty and fidelity by being transformed into trees.[24] In the face of impending death Baucis and Philemon were granted further life, being transformed into an oak and a lime tree, which Ovid relates were treated as if they were the gods themselves.

In contrast to the rewarding of Baucis and Philemon, many cases of transformation from human to plant have an air of punishment, a never-ending, purgatorial banishment from the human realm. In a number of Greek sources, the nymph Pitys is transformed into a pine tree as a punishment for refusing marriage.[25] The transformation of Cyparissus into the cypress tree following the accidental killing of his beloved pet stag is very similar to the myth of the Heliades. Having begged Apollo that he may mourn forever, Cyparissus is told, "You shall be mourned by me, shall mourn for others, and your place shall be always where others grieve."[26] Yet the cypress is more than just a symbol of human grief. The cypress expresses the pains of sentient plant life.

In the myth of Narcissus, the story of the beautiful Greek youth who fell in love with his own reflection, it is not punishment, but relief and release that permeate the story. Ovid narrates that after he had glimpsed his own reflection, the youth sat transfixed, anchored to the spot at the side of a pool, gazing at his own beauty, slowly wasting away. After he had finally passed away, all the Nymphs mourned for Narcissus, but as they were about to put his body on a funeral pyre, "In place of his body they find a flower, its yellow centre girt with white petals."[27] Narcissus had been transformed into a flower. Other versions of this myth tell how Narcissus killed himself and from his blood sprang the golden flower.

In contrast to Marder, who sees only anthropomorphic self-projection of the human upon nature in the myth of Narcissus, I regard these myriad transformations between the human and plant realm in the Greek myths as recognition of the continuities between the worlds of the human and plant. Humans and plants share in our basic substance and those most important markers of earthly life, birth and death. These significant and substantial connections are made real through the repeated passing away of human beings and the subsequent coming into existence of plant forms (and vice versa).[28]

These profound connections with the plant kingdom are by no means restricted to ancient Greece. The Ancient Egyptian *Tale of Two Brothers* tells the story of one brother, Bata, placing his soul high up in an acacia tree. This transmigration of the soul leads the human Bata to a vulnerable state, for when the tree is cut down (as, sadly, trees often are), he dies instantly.

In the myth of the origins of the kava plant (*Piper methysticum*), used as a ritual psychoactive and sedative in the South Pacific, the plant is connected to the death of an ancestral being, in a similar fashion to the deaths of Adonis and Minthe. In the stories collected in Tonga by Edward Gifford, kava originates from the death of the daughter of Fevanga and Fefafa, who, in one variant, is called Kavaonau.[29] According to a version from the Ha'apai group of islands:

> For five nights Fevanga and Fefafa kept visiting the grave of their daughter, and after five nights there was growing from her head a kava plant, and from her guts there grew a sugar cane. That kava grew large, and the cane grew also.[30]

This metamorphosis from human to plant form is echoed in a number of other Pacific tales, including the story of the goddess *Lau-ka-'ie'ie* (leaf of the trailing pandanus), who on death transformed into the fiery *'ie'ie* vine (*Freycinetia arborea*).[31]

In the Indian spiritual traditions, many plants are described as originating from the bodies of deities (See chapter 2). Foundational texts of the Vedic tradition, such as the *Chāndogya Upaniṣad*, relate a fundamental sense of connection and relatedness permeating all the phenomena in the natural world from humans to animals to plants and water and rocks—"Brahman, you see, is this whole world."[32] In the *Mahābhārata* it is recognized that the worlds of plants, humans, and animals interpenetrate. As in ancient Greece, they do so through a process of death and rebirth—as the *Mahābhārata* states, "All creatures, stupefied, in a consequence of ignorance, by the attributes of goodness and passion and darkness are continually revolving like a wheel."[33]

The description of the cycle of reincarnation and the understanding that humans may be reborn as plants (and vice versa) first appears in early *Upaniṣads*. The *Brahdāranyaka Upaniṣad* describes the concept of *karma* when the sage Yajnavalkya is asked by the scholar Artabhaga what happens to a person after they have died.[34] The sage describes two paths that can be taken, the path that leads to a final existence or the path that leads to a rebirth in the world. The path of rebirth into the world includes the possibility of humans being transformed into vegetation via the smoke of a funeral pyre.[35]

The potential to be reborn as a plant also appears in a passage in the *Chāndogya Upaniṣad*. Interestingly, the transmigration of the soul is described entirely in *physical* terms. Unlike in a Cartesian world, there is no division between the movement of matter and the movement of mind or soul. As in the previous account, the dead person is again transformed into smoke by the funeral flames; the smoke again merges with the rainclouds and falls back to earth to become trees and crops.[36] The cycle begins again when human beings consume the rice and barley—plants become humans once more. In a similar fashion to the myths of ancient Greece, it is acknowledged that through the transformation of matter it is possible for a plant to be reborn as a human being and for a human being to be reborn as a plant.

Many Indigenous cultures hold similar stories. In Aboriginal Australia, a basic sense of kinship between human beings and plants arises from the stories of the Dreaming beings—the creative ancestral beings that shaped the Australian landscape into its current form. The Dreaming beings created the relationships, and the rules of those relationships, which are known to Aboriginal people as Law. As Hobbles Danyari says:

> Everything come up out of ground—language, people, emu, kangaroo, grass. That's Law.[37]

The species that exist in country have their origins in the Dreaming beings, many of which transformed from an initial humanity into the forms (plants, animals, rocks) in which they exist today. The Adnyamathanha people in the Flinders Ranges tell of a metamorphosis concerning the Iga tree (*Capparis mitchelii*). The Iga was initially a Dreaming being who traveled in the shape of a man, heading south to find a woman in the country of the Adyamathanha. As they were digging for *ngarndi* roots, the man and woman were startled by something nearby and said:

> "Well, look, if we turn into a tree, they mightn't take any notice of us."
> So they came along as trees from there.[38]

The Igas transformed from humans into trees, creating the Iga species. This attribution of an initial humanity to plant species is a way of representing the subjective nature of the beings in question. The personalization of features in the natural landscape is not a primitive projection of human characteristics, but a sophisticated method "for encountering the world."[39] Interestingly, the myth of the Iga contains both a recognition of personhood, and an acknowledgment that trees can often be poorly attended to.

In Australia, there are many stories of metamorphosis to add to those of the Iga. One Gunwinggu tale from the Northern Territory tells of the couple Namalbi and Ngalmadbi, who left their camp after a quarrel with their family and found themselves changing into the Pandanus trees which populate the country.[40] In another Gunwinggu story, the old man Mananda, who was unable to walk very far, one day told his sons that he would remain in one spot while they went off, and, "He just sat there for so long that he became a long yam."[41] The Ngulugwongga of the Daly River in the far north of Australia tell a similar tale. A Dreaming woman, Yilig-moi-indih or "Red Lily Woman" came from another country carrying the roots of the red lily (*Nelumbo nucifera*). She planted the roots in the ground along her journey and after she had finished her planting, she sat down on the ground. At that spot she transformed into the red lily root and went down into the ground where she is today.[42] In the Manjindji country of the central desert region in Western Australia, two men, (Wadi Gudjara), traveled from west to southeast. After coming into contact with a camp of people that mysteriously disappeared, they stamped on the ground where the people had sat and yams sprang up in their place. The teller of the story, Joe Mungu, relates that "these vegetables were once men, women, girls and boys—until the Wadi Gudjara came along."[43]

The repeated metamorphosis between human beings, plants, and animals greatly overlaps with ideas of kinship and shared lineage explored in chapter 1. To a large extent, the sharing of substance through transformation creates a "mutuality of being" that is the marker of kinship.

For the Koyukon people of Alaska, all things on Earth were created in a mythical time known as Kk'adonts'idnee, which Richard Nelson translates as Distant Time.[44]

> During this age "the animals were human"—that is, they had human form, they lived in human society, and they spoke human (Koyukon) language. At some point in the Distant Time, certain humans died and were transformed into animal or plant beings, the species that inhabit Koyukon country today.[45]

In this myth, there is yet another transformation from human into plant, a transmutation of the human body into plant flesh at the time of death. There is also a portrayal of common ancestry and the sharing of a fundamental substance. At the end of the mythological age, a flood covered the Earth, all human beings were killed, and Raven had to make them anew, in the form they are today. The Tlingit (neighbors of the Koyukon) say that Raven made these new people out of wood and leaves, which is why humans are now mortal.[46] Once again, plants feature at the birth of humankind and humans naturally share mortality with the trees from which they were made.

METAMORPHOSIS

FROM TREES AND FLOWERS

"You, corn; you, tzité; you, fate; you, creature; get together, take each other," they said to the corn, to the tzité, to fate, to the creature. "Come to sacrifice here, 'heart of heaven'; do not punish Tepeu and Gucumatz!"

Then they talked and spoke the truth: "Your figures of wood shall come out well; they shall speak and talk on earth."

"So may it be," they answered when they spoke.

And instantly the figures were made of wood. They looked like men, talked like men, and they populated the surface of the earth.

They existed and multiplied; they had daughters, they had sons, these wooden figures; but they did not have souls, nor minds, they did not remember their creator, their maker; they simply walked on all fours, aimlessly.

They no longer remembered the "heart of heaven" and therefore they fell out of favour. It was merely a trial, an attempt at man. At first they spoke, but their face was without expression; their feet and hands had no strength; they had no blood, nor substance, nor moisture, nor flesh; their cheeks were dry, their feet and hands were dry, and their flesh was yellow.

Therefore, they no longer thought of their creator nor their maker, nor of those who made them and cared for them. These were the first men who existed in great numbers on the face of the Earth. (*Popol Vuh*)[47]

"Well," said Math, "we will seek, I and thou, by charms and illusion, to form a wife for him out of flowers. He has now come to man's stature, and he is the comeliest youth that was ever beheld." So they took the blossoms of the oak, and the blossoms of the broom, and the blossoms of the meadowsweet, and produced from them a maiden, the fairest and most graceful that man ever saw. And they baptized her, and gave her the name of Blodeuwedd. (*Four Branches of the Mabinogi*)[48]

The Creator began to raise up the people and nations that are in that region, making one of each nation of clay, and painting the dresses that each one was to wear. Those that were to wear their hair, with hair; and those that were to be shorn, with hair cut; and to each nation was given the language that was to be spoken, and the songs to be sung, and the seeds and food that they were to sow. When the Creator had finished painting and making the said nations and figures of clay, he gave life and soul to each one, as well men as women, and ordered that they should pass under the earth. Thence each nation came up in the places to which he ordered them to go. Thus they say that some came out of caves, others issued from hills, others from fountains, others from the trunks of trees. From this cause, and owing to having come forth and commenced to multiply, from those places, and to having had the beginning of their lineage in them, they made huacas and places of worship of them in memory of the origin of their lineage which proceeded from them. Thus each nation uses the dress with which they invest their huaca; and they say that the first that was born from that place was there turned into stones, others say the first of their lineages were turned into falcons, condors, and other animals and birds. Hence the huacas they use and worship are in different shapes. (*The Fables and Rites of the Incas*)[49]

BENEDETO
MONTAGNA

Some god did listen to her prayer; her
last petition had its answering gods. For even as she
spoke the earth closed over her legs; roots burst
forth from her toes and stretched out on either side
the supports of the high trunk; her bones gained
strength, and, while the central pith remained the
same, her blood changed to sap, her arms to long
branches, her fingers to twigs, her skin to hard bark.
And now the growing tree had closely bound her
heavy womb, had buried her breast and was just
covering her neck; but she could not endure the
delay and, meeting the rising wood, she sank down
and plunged her face in the bark. Though she has
lost her old-time feelings with her body, still she
weeps, and the warm drops trickle down from the
tree. Even the tears have fame, and the myrrh which
distils from the tree-trunk keeps the name of its mistress and will be remembered through all the ages.

But the misbegotten child had grown within the
wood, and was now seeking a way by which it might
leave its mother and come forth. The pregnant tree
swells in mid-trunk, the weight within straining on
its mother. The birth-pangs cannot voice themselves, nor can Lucina be called upon in the words
of one in travail. Still, like a woman in agony, the
tree bends itself, groans oft, and is wet with falling tears. Pitying Lucina stood near the groaning
branches, laid her hands on them, and uttered
charms to aid the birth. Then the tree cracked
open, the bark was rent asunder, and it gave forth
its living burden, a wailing baby-boy. The naiads
laid him on soft leaves and anointed him with his
mother's tears. (Ovid, *Metamorphoses*)[50]

After the term of his silence was over he also visited the great Antioch, and passed into the Temple of the Apollo of Daphne, to which the Assyrians attached the legend of Arcadia. For they say that Daphne, the daughter of Ladon, there underwent her metamorphosis, and they have a river flowing there, the Ladon, and a laurel tree is worshipped by them which they say was substituted for the maiden; and cypress trees of enormous height surround the Temple, and the ground sends up springs both ample and placid, in which they say Apollo purified himself by ablution. And there it is that the earth sends up a shoot of cypress, they say in honour of Cyparissus, an Assyrian youth; and the beauty of the shrub lends credence to the story of his metamorphosis. Well, perhaps I may seem to have fallen into a somewhat juvenile vein to approach my story by such legendary particulars as these, but my interest is not really in mythology. What then is the purport of my narrative? Apollonius, when he beheld a Temple so graceful and yet the home of no serious studies, but only of men half-barbarous and uncultivated, remarked: "O Apollo, change these dumb dogs into trees, so that at least as cypresses they may become vocal." (Philostratus, *Life of Apollonius of Tyana*)[51]

This is how the story of Daphne, the daughter of Amyclas, is related. She used never to come down into the town, nor consort with the other maidens; but she got together a large pack of hounds and used to hunt, either in Laconia, or sometimes going into the other countries of the Peloponnese. For this reason she was very dear to Artemis who gave her the gift of shooting straight.

On one occasion she was traversing the country of Elis, and there Leucippus, the son of Oenomaus, fell in love with her; he resolved not to woo her in any common way, but assumed women's clothes, and in the guise of a maiden, joined her hunt. And it so happened that she very soon became extremely fond of him, nor would she let him quit her side embracing him and clinging to him at all times.

But Apollo was also fired with love for the girl, and it was with feelings of anger and jealousy that he saw Leucippus always with her; he therefore put it into her mind to visit a stream with her attendant maidens, and there to bathe. On their arrival there, they all began to strip; and when they saw that Leucippus was unwilling to follow their example, they tore his clothes from him: but when they thus became aware of the deceit he had practised and the plot he had devised against them, they all plunged their spears into his body. He, by the will of the gods, disappeared; but Daphne, seeing Apollo advancing upon her, took vigorously to flight; then as he pursued her, she implored Zeus that she might be translated away from mortal sight, and she is supposed to have become the bay-tree which is called *daphne* after her. (Parthenius, *Love Romances*)[52]

Thus the goddess warned and through the air,
drawn by her swans, she took her way; but the
boy's manly courage would not brook advice. It
chanced his hounds, following a well-marked trail,
roused up a wild boar from his hiding-place; and, as
he was rushing from the wood, the young grandson
of Cinyras pierced him with a glancing blow.
Straightway the fierce boar with his curved snout
rooted out the spear wet with his blood, and pursued
the youth, now full of fear and running for his life;
deep in the groin he sank his long tusks, and
stretched the dying boy upon the yellow sand.
Borne through the middle air by flying swans on
her light car, Cytherea [Aphrodite] had not yet come to Cyprus,
when she heard afar the groans of the dying youth
and turned her white swans to go to him. And
when from the high air she saw him lying lifeless
and weltering in his blood, she leaped down, tore
both her garments and her hair and beat her breasts
with cruel hands. Reproaching fate, she said: "But
all shall not be in your power. My grief, Adonis,
shall have an enduring monument, and each passing
year in memory of your death shall give an imitation
of my grief. But your blood shall be changed to a
flower. Or was it once allowed to thee, Persephone,
to change a maiden's [the nymph Menthe] form to fragrant mint, and
shall the change of my hero, offspring of Cinyras, be
grudged to me?" So saying, with sweet-scented
nectar she sprinkled the blood; and this, touched
by the nectar, swelled as when clear bubbles rise up
from yellow mud. With no longer than an hour's
delay a flower sprang up of blood-red hue such as
pomegranates bear which hide their seeds beneath
the tenacious rind. But short-lived is their flower;
for the winds from which it takes its name [Anemone] shake
off the flower so delicately clinging and doomed too
easily to fall. (Ovid, *Metamorphoses*)[55]

On the learned leaves of Apollon's mournful Iris was embroidered many a plant-grown word; and when Zephyros breathed through the flowery garden, Apollon turned a quick eye upon his young darling, his yearning never satisfied; if he saw the plant beaten by the breezes, he remembered the quoit, and trembled for fear the wind, so jealous once about the boy, might hate him even in a leaf: if it is true that Apollon once wept with those eyes that never wept, to see that boy writhing in the dust, and the pattern there on the flower traced its own "alas!" on the Iris, and so figured the tears of Phoibos. (Nonnus, *Dionysiaca*)[53]

"I am the author of thy death.
And yet, what is my fault, unless my playing with
thee can be called a fault, unless my loving thee
can be called a fault? And oh, that I might
give up my life for thee, so well-deserving, or give
it up with thee! But since we are held from this
by the laws of fate, thou shalt be always with me, and
shalt stay on my mindful lips. Thee shall my lyre,
struck by my hand, thee shall my songs proclaim. And
as a new flower, by thy markings shalt thou imitate
my groans. Also the time will come when a most
valiant hero [Ajax] shall be linked with this flower, and
by the same markings shall he be known." While
Apollo thus spoke with truth-telling lips, behold, the
blood, which had poured out on the ground and
stained the grass, ceased to be blood, and in its place
there sprang a flower brighter than Tyrian dye. It
took the form of the lily, save that the one was of
purple hue, while the other was silvery white. Phoebus, not satisfied with this—for 'twas he who wrought
the honouring miracle—himself inscribed his grieving
words upon the leaves, and the flower bore the marks,
AI AI, letters of lamentation, drawn thereon. Sparta,
too, was proud that Hyacinthus was her son, and even
to this day his honour still endures; and still, as the
anniversary returns, as did their sires, they celebrate
the Hyacinthia in solemn festival. (Ovid, *Metamorphoses*)[54]

Zeus smote Phaethon with a thunderbolt and brought back the sun to its accustomed course. And Phaethon fell to the earth at the mouths of the river which is now known as the Pados (the river Po), but in ancient times was called the Eridanos, and his sisters vied with each other in bewailing his death and by reason of their exceeding grief underwent a metamorphosis of their nature, becoming poplar trees.

And these poplars, at the same season each year, drip tears (or sap), and these, when they harden, for what men call amber, which in brilliance excels all else of the same nature and is commonly used in connection with the mourning attending the death of young. (Diodorus Siculus, *Library of History*)[56]

Mink man went to several tree-women and told them that their husband (Raven) had been killed:

When one woman heard the story, she cried and pinched her skin. Then she was changed into the spruce tree, with its rough and pinched bark.

When another heard it, she cried and slit her skin with a knife. She became a poplar, with its deeply cut bark.

When a third woman was told the story, she cried and pinched herself until she bled. She turned into the alder, whose bark is used to make red dye. (Koyukon Myth)[57]

When he had spoken a word with Baucis, Philemon announced their joint decision to the gods: "We ask that we may be your priests, and guard your temple: and, since we have spent our lives in constant company, we pray that the same hour may bring death to both of us—that I may never see my wife's tomb, nor be buried by her." Their request was granted.

They had the care of the temple as long as they lived. And at last, when, spent with extreme old age, they chanced to stand before the sacred edifice talking of old times, Baucis saw Philemon putting forth leaves, Philemon saw Baucis; and as the tree-top formed over their two faces, while still they could they cried with the same words: "Farewell, dear mate," just as the bark closed over and hid their lips. Even to this day the Bithynian peasant in that region points out two trees standing close together, and growing from one double trunk. These things were told me by staid old men who could have had no reason to deceive. With my own eyes I saw votive wreaths hanging from the boughs, and placing fresh wreaths there myself, I said: "Those whom the gods care for are gods; let those who have worshipped be worshipped." (Ovid, *Metamorphoses*)[58]

CYPARISSUS

Amidst this throng came the cone-shaped cypress,
now a tree, but once a boy, beloved by that god who
strings the lyre and strings the bow. For there was
a mighty stag, sacred to the nymphs who haunt the
Carthaean plains, whose wide-spreading antlers gave
ample shade to his own head. His antlers gleamed
with gold, and down on his shoulders hung a gem-
mounted collar set on his rounded neck. Upon his
forehead a silver boss bound with small thongs was
worn, and worn there from his birth. Pendent from
both his ears, about his hollow temples, were gleaming pearls. He, quite devoid of fear and with none
of his natural shyness, frequented men's homes and
let even strangers stroke his neck. But more than
to all the rest, O Cyparissus, loveliest of the Cean
race, was he dear to you. 'Twas you who led the
stag to fresh pasturage and to the waters of the clear
spring. Now would you weave bright garlands for
his horn; now, sitting like a horseman on his back,
now here, now there, would gleefully guide his soft
mouth with purple reins.

'Twas high noon on a summer's day, when the
spreading claws of the shore-loving Crab were burn-
ing with the sun's hot rays. Weary, the stag had
lain down upon the grassy earth and was drinking in
the coolness of the forest shade. Him, all unwit-
tingly, the boy, Cyparissus, pierced with a sharp
javelin, and when he saw him dying of the cruel wound,
he resolved on death himself. What did not Phoebus
say to comfort him! How he warned him to grieve
in moderation and consistently with the occasion!
The lad only groaned and begged this as the
boon he most desired from heaven, that he might

mourn for ever. And now, as his life forces were
exhausted by endless weeping, his limbs began to
change to a green colour, and his locks, which but
now overhung his snowy brow, were turned to a
bristling crest, and he became a stiff tree with slender
top looking to the starry heavens. The god groaned
and, full of sadness, said: "You shall be mourned
by me, shall mourn for others, and your place shall
always be where others grieve." (Ovid, Metamorphoses)[59]

And this is what shall come to pass, that I shall draw out my soul, and I shall put it upon the top of the flowers of the acacia, and when the acacia is cut down, and it falls to the ground, and thou comest to seek for it, if thou searchest for it seven years do not let thy heart be wearied. For thou wilt find it, and thou must put it in a cup of cold water, and expect that I shall live again, that I may make answer to what has been done wrong…

Now many days after these things, the younger brother was in the valley of the acacia; there was none with him; he spent his time in hunting the beasts of the desert, and he came back in the even to lie down under the acacia, which bore his soul upon the topmost flower. And after this he built himself a tower with his own hands, in the valley of the acacia; it was full of all good things, that he might provide for himself a home.

And he went out from his tower, and he met the Nine Gods, who were walking forth to look upon the whole land. The Nine Gods talked one with another, and they said unto him, "Ho! Bata, bull of the Nine Gods, art thou remaining alone?"… And Ra Harakhti said to Khnumu, "Behold, frame thou a woman for Bata, that he may not remain alive alone." And Khnumu made for him a mate to dwell with him. She was more beautiful in her limbs than any woman who is in the whole land…

And Bata loved her very exceedingly, and she dwelt in his house; he passed his time in hunting the beasts of the desert, and brought and laid them before her. He said, "Go not outside, lest the sea seize thee; for I cannot rescue thee from it, for I am a woman like thee; my soul is placed on the head of the flower of the acacia; and if another find it, I must fight with him." And he opened unto her his heart in all its nature.

Now after these things Bata went to hunt in his daily manner. And the young girl went to walk under the acacia which was by the side of her house. Then the sea saw her, and cast its waves up after her. She betook herself to flee from before it. She entered her house. And the sea called unto the acacia, saying, "Oh, would that I could seize her!" And the acacia brought a lock from her hair, and the sea carried it to Egypt…

And many days after these things the people who were sent to strange lands came to give report unto the king: but there came not those who went to the valley of the acacia, for Bata had slain them, but let one of them return to give a report to the king. His majesty sent many men and soldiers, as well as horsemen, to bring her back. And there was a woman amongst them, and to her had been given in her hand beautiful ornaments of a woman. And the girl came back with her, and they rejoiced over her in the whole land.

And his majesty loved her exceedingly, and raised her to high estate; and he spake unto her that she should tell him concerning her husband. And she said, "Let the acacia be cut down, and let one chop it up." And they sent men and soldiers with their weapons to cut down the acacia; and they came to the acacia, and they cut the flower upon which was the soul of Bata, and he fell dead suddenly. (*Tale of Two Brothers*)[61]

Mimosa arabica.

Narcissus Poeticus   Narcisse l. P.

So he [Bacchos] spoke to the breezes of spring, while walking in a flowery meadow. Beside a fragrant myrtle he stayed his feet for a soothing rest at mid-day. He leaned against a tree and listened to the west breeze whispering, overcome by fatigue and love; and as he sat there, a Hamadryas Nymph at home in the clusters of her native tree, a maiden unveiled, peeped out and said, true both to Cypris and to loving Lyaios:

"Bacchos can never lead Aura to his bed, unless he binds her first in heavy galling fetters, and winds the bonds of Cypris round hands and feet: or else puts her under the yoke of marriage in sleep and steals the girl's maidenhood without brideprice."

Having spoken she hid again in the tree her agemate, and entered her woody home…

While Bacchos would be preparing a cunning device for her bed, Lelantos's daughter wandered about seeking a fountain, for she was possessed with parching thirst. Dionysos failed not to see how thirsting Aura ran rapidly over the hills. Quickly he leapt up and dug the earth with his wand at the foundation of a rock: the hill parted, and poured out of itself a purple stream of wine from its sweet scented bosom. The Seasons, Handmaids of Helios, to do grave to Lyaios, painted with flowers the fountain's margin, and fragrant whiffs of the new-growing meadow beat on the balmy air. There were the clustering blooms which have the name of Narcissos the fair youth, whom horned Selene's bridegroom Endymion begat on leafy Latmos, Narcissos who long ago gazed on his own image formed in the water, that dumb image of a beautiful deceiver, and died as he gazed on the shadowy phantom of his shape: there was the living plant of the Amyclaian iris [Hyacinthos once more!]; there sang the nightingales over the spring blossoms, flying in troops above the clustering flowers.

There were the clustering blooms which have the name Narkissos the fair youth, whom horned Selene's bridegroom Endymion begat on leafy Latmos. (Nonnus, *Dionysiaca*)[60]

This is the story of how kava grew.

It is said that there was once a chief called Loau, whose ancestors resided in Lifuka, and for whom the district of Haaloau in Lifuka is named. It is said that his dwelling had eight enclosures or fences and that a great number of people lived there.

Whilst Loau resided at Haamea, a man called Fevanga paid a visit to Loau. The name of Fevanga's wife was Fefafa. After residing some time with Loau, Fevanga told him that he would like to go to Eueiki to see his relatives and that he would soon return again. To this the chief agreed. Fevanga went to the island of Eueiki and stopped there with his wife. They had a daughter who was a leper. Time went on and still Fevanga tarried in Eueiki. Loau missed Fevanga and finally decided to go to Eueiki himself, so he had his dependents prepare for the voyage. A large rowing canoe (tafaanga) was launched and away they went to Eueiki. They arrived there at dusk. Loau ordered that the canoe be carried to Fevanga's home and put close to a large *kape* plant (*Arum costatum*), with the outrigger on top of the *kape*.

Fevanga came down to greet his visitors and they responded, saying: "Happy to see you in good health in this island." Loau sat down with his back to the big *kape*, whilst Fevanga searched for food. Fevanga's search was not fruitful, for Eueiki was suffering from famine at the time. Nevertheless, he fired his earth oven and at the same time suggested to Loau that, if he would not mind going down to the beach, he would find it cooler there. Fevanga was desirous that Loau should move in order that he might dig up the *kape* plant to roast.

After Loau had accommodatingly removed to the beach, Fevanga dug up the big *kape* plant and put it in the oven. He then killed his leprous daughter and roasted her together with the *kape*. Shortly after Loau and his men returned, the oven was opened, and the food set before Loau. Loau issued orders that the head of Fevanga's unfortunate daughter be cut off and buried in one place, while the body was to be buried in another place. Loau told Fevanga to take notice that two plants would grow from the head and that he was to care for them. Farewells were said and Loau returned to Tongatabu.

Fevanga remained in Eueiki to care for the plants, as it was his duty to take them to Loau in Haamea when they had reached maturity. They proved to be kava and sugar cane. He watched them carefully and, one day when they were nearly full grown, he saw a rat gnawing the kava. After eating the kava, the rat chewed the sugar cane. All the Tongan people drink the kava and eat the sugar cane, because the rat ate the kava first and then the sugar cane. Then Fevanga knew that the time had arrived to pull up the two plants and take them to Tongatabu for a meeting of the chiefs.

When Loau saw Fevanga approaching with the plants he cried: "This is the kava of Fevanga and Fefafa from Faimata. A single chief for the *olovaha* (i.e., the plain under side of the kava bowl which is towards the presiding chief at a kava party), and many for the *apaapa* (the place occupied by other chiefs at a kava party). Husk of the coconut for cleaning the kava root." A bowl was brought and a *matapule* directed a person from the *toua* (the place occupied in a kava party by the people as opposed to the *alofi*, the place of the chiefs) to make kava. Coconut husks were used to gather the pieces of kava in, as it was split. Then it was given to the people sitting in the *toua* to be chewed. After being chewed, it was placed in the bowl, mixed, and served. Directions were issued to chop the sugar cane, which was used as a relish (the yam, banana, or other food eaten at a kava drinking ceremony) with the kava.

The place where the kava grew is still to be seen in Eueiki even unto this day. (Told by Malaki Lavulo of Pangai, Lifuka Island, Ha'apai)[62]

Tom. VIII.
Tab. 64.

"Yâgñavalkya," he said, "when such a person (a sage) dies, do the vital breaths (prânas) move out of him or no?"

"No," replied Yâgñavalkya; "they are gathered up in him, he swells, he is inflated, and thus inflated the dead lies at rest."

"Yâgñavalkya," he said, "when such a man dies, what does not leave him?"

"The name," he replied; "for the name is endless, the Visvedevas are endless, and by it he gains the endless world."

"Yâgñavalkya," he said, "when the speech of this dead person enters into the fire, breath into the air, the eye into the sun, the mind into the moon, the hearing into space, into the earth the body, into the ether the self, into the shrubs the hairs of the body, into the trees the hairs of the head, when the blood and the seed are deposited in the water, where is then that person?"

Yâgñavalkya said: "Take my hand, my friend. We two alone shall know of this; let this question of ours not be (discussed) in public." Then these two went out and argued, and what they said was karman (work), what they praised was karman, viz. that a man becomes good by good work, and bad by bad work. (*Brhadāranyaka Upaniṣad*)[63]

Having dwelt there, till their (good) works are consumed, they return again that way as they came, to the ether, from the ether to the air. Then the sacrificer, having become air, becomes smoke, having become smoke, he becomes mist.

Having become mist, he becomes a cloud, having become a cloud, he rains down. Then he is born as rice and corn, herbs and trees, sesamum. and beans. From thence the escape is beset with most difficulties. For whoever the persons may be that eat the food, and beget offspring, he henceforth becomes like unto them.

Those whose conduct has been good, will quickly attain some good birth, the birth of a Brâhmana, or a Kshatriya, or a Vaisya. But those whose conduct has been evil, will quickly attain an evil birth, the birth of a dog, or a hog, or a Kandâla.

On neither of these two ways those small creatures (flies, worms, &c.) are continually returning of whom it may be said, Live and die. (*Chandogya Upaniṣad*)[64]

There was a single man whom came down from Yaramangga—the Iga man. When he got a bit further down the river to Nakatangga, to a waterhole there called Paddy Paddy, between Nakatangga and Narialka he picked up his woman.

They went on down the river from there. At the junction of the Wilson and Cooper Rivers, not far from Nappa Merrie, they were digging *ngarndi*. That's why there's a big lake there, between the two rivers. The water always lies there—where the Iga man and woman were digging.

They were frightened away from there—something hunted them out. They said: "Well, look, if we turn into a tree, they mightn't take any notice of us." So they came along as trees from there …

An argument sprang up at a place called Mulga something. They had a big argument there. The other trees got stuck into them. They had a big fight. The pine trees, and all the mulgas, and the coolabah—everybody got together.… So they thought: "Oh well, we'll go elsewhere"…

The biggest camp they had was at Nipa Awi on Wooltana. They camped there for a while. There's a big mob of Iga trees there, with a mob of little ones where the old camp was.

From there they came down to Ithala Awi. At Ithala Awi they mucked around for a while, about a month. Then another argument started up. There are Iga trees everywhere there … (Adnyamathanha Dreaming)[65]

Mananda was an old man, unable to walk far: "My sons," he said, "I have to stay here. You go and find the places and come back and tell me about them." He sat down there and made himself like a long yam: "He just sat there for such a long time that he became a long yam." (Gunwinggu Dreaming)[66]

The two were getting ready "to make themselves," with "horns" rising from their heads, and stood there, looking back at the country from which they had come. When they saw the others they began to run. Raramin and Djuwulgid and the child tried to catch up with them, to bring them back. But the other two said: "It's too late! We have made ourselves pandanus; we can't come back now." The "horns" as pandanus shoots, rose from their heads, and they became pandanus trees. (Gunwinngu Dreaming)[67]

Yilig-moi-indiji, Red Lily woman, came from another country carrying a basket full of lily roots suspended from her head by a fibre cord. She came to wag-Bamayag, where she put in a lily root. At deg-Dugailyil she put another, and then several at deg-Nangad.... Later she picked up her baster of lily roots and went over to the grass, cleaned an area and put some lily roots there, covering them with a little mud, just a little way below the surface of the ground. However Braulum and the other women had followed her, and Braulum asked why she had not put them deeper in the ground. Yilig replied, "No, not deep, but near the surface." When Yilig had gone, Braulum put some of them deep into the ground ["so that you have to dig deeply to get these roots"] but she also collected some for eating.

Yilig picked up her basket and went on to Bundjarang-ginya on the other side of the Creek, where she sat down.

Braulum and the other women were meanwhile roasting the lily roots at Bandirang. "Come, the roots are roasted," they called. "They are ready to take out." They began to remove them, and were amazed to find that there was nothing there—only stones? All the roots had gone underground to wag-Bundjarang-ginya, near the river where Yilig was sitting. There she remains, Dreaming, with her basket as a rock. (Mandjindji Dreaming)[68]

Would curses kill,
as doth the mandrake's groan,
I would invent as bitter-searching terms,
As curst, as harsh and horrible to hear,
—*Henry VI, Part 2*

4

LEGEND

Sir Thomas Browne, was, for want of a better term, a "science communicator" in seventeenth-century England—a time of cholera, Oliver Cromwell, and Anglo-Dutch warfare. His most famous work *Pseudodoxia Epidemica* or *Enquiries Into Very Many Received Tenets and Commonly Presumed Truths* was a pioneering investigation into a whole array of the superstitious practices of the day, from the popular understanding of unicorns to the presumed evils of sitting cross-legged. First published in 1658, six years before New York would shed its old name of New Amsterdam, the popular book was a champion of the Baconian scientific method, and was one of the first texts that began to debunk widely understood beliefs about the natures of common plants. In the sixth chapter of *Pseudodoxia*, Browne provides rational alternatives to superstitions surrounding a number of plants, including mistletoe and the rose of Jericho, cinnamon and almonds, but most of the chapter is reserved for his discussions of one of the most mysterious plants, the mandrake (*Mandragora officinarum*).

With a mixture of empirical observation and common sense, Browne demolishes superstitions involving mandrake. One of these told of the mandrake growing up in places of execution from "fat or urine that drops from the body of the dead."[1] Another was that the mandrake gave a shriek when pulled from the ground, which Browne dismissed as "ridiculous," pointing to a small noise which the root made when pulled from the ground as the source of the concern.[2]

> The third affirmeth the roots of Mandrakes do make a noise, or give a shriek upon eradication; which is indeed ridiculous, and false below confute; arising perhaps from a small and stridulous noise, which being firmly rooted, it maketh upon divulsion of parts.[3]

Browne also dismisses similar accounts of the root moly in Homer and the root Baaras described by Josephus with a withering description of "their mutual concurrences supporting their solitary instabilities."[4]

To a modern reader, all of the preceding excerpts, from the tales of the sacred oaks to the travels of the Iga tree, could easily be treated in such a fashion, labeled as a superstitious legend, unable to withstand the withering gaze of Browne's Baconian empiricism. To a skeptic, all of the stories in this volume could be labeled as "myths" in the sense of their being widely held but "untrue" legends. This is both a common and an easy stance to take.

However, when comparing across cultures, it is prudent not to treat all myths as widely held misconceptions. The origin stories of Aboriginal Australia are not equivalent to medieval superstitions concerning the ability of goat blood to soften diamond or the ability of the plant moonwort (*Lunaria*) to break locks or to draw the shoes off horses. Thomas Browne himself makes this distinction during his discussion of the hieroglyphs of ancient Egypt, in a later chapter in *Pseudodoxia*. Browne points out the fact that in depicting gods and goddesses composed of a compendium of bodily parts, the Ancient Egyptians "took a liberty to compound and piece together creatures of allowable forms into mixtures inexistent" doing so to "express complexed significations."[5] Browne separates these knowledgeable, allegorical personifications of earthly life from later depictions by medieval writers who literally believed in the existence of creatures such as Griffins and Phoenixes. As Browne states, clearly these are

> [p]ieces of good and allowable invention unto the prudent Spectator, but are lookt on by vulgar eyes as literal truths, or absurd impossibilities; whereas indeed, they are commendable inventions, and of laudable significations.[6]

Taking our cues from Thomas Browne, a distinction can be made between legends and myths. Botanical legends are perhaps best described as the inauthentic stories of fantastical plant natures and characteristics from faraway places. Whereas, the botanical myths that we have seen are local stories that use personifications and nonliteral, allegorical portrayals of earthly life, in part to reveal the relationships between humans and the natural world. However, while the former category is largely composed of credulous, literal explanations, they can also include aspects of Browne's "laudable significations." The categories of "legend" and "myth" are therefore not necessarily discrete. Legend and myth often overlap, and when they do, this is another rich source of stories about the relationships between human beings and plants.

SOME PART OF TRUTH: THE BARNACLE AND THE WAK-WAK

Like the myths presented so far, botanical legends are many and varied, but common themes nevertheless emerge. These are what Thomas Browne calls the "some part of truth" that he considered to be buried in many of the superstitions of his time. Echoing those botanical myths presented in chapter 3, a common theme in these legends is that of metamorphosis, the transformation of forms between human and plant, between plant and animal, and even between plant and plant.

One of the most famous legends concerning the metamorphosis of plants is that of the Barnacle Tree, a legendary tree described by the sixteenth-century, early-Renaissance herbalist John Gerard. The Barnacle Tree was a tree

> upon which grow certain shell fishes, of a white colour, tending to russet; wherein are contained little living creatures, which shells in time of maturity do open, and out of them grow those little living things; which falling into the water, do become fowls, which we call Barnacles, in the north of England brant geese and in Lancashire tree geese.[7]

According to Gerard, the Barnacle was a tree whose fruits metamorphosed into geese. This is an idea that echoes that of the transformation of species from one kind to another, an idea which is found in Pliny's *Natural History*. Its inclusion in the *National History* is significant as this was a text that "became a substitute for a general education" in Medieval Europe following the loss of the more empirical Greek texts of Theophrastus, the founding father of botany.[8] In the *Natural History* Pliny presents an account of the transformation of

> a plane-tree at Laodicea changed into an olive on the arrival of Xerxes. Not to launch out into an absolutely boundless subject, the volume by Aristander teems with portents of this nature in Greece, as do the Notes of Gaius Epidius in our own country, including cases of trees that talked.[9]

In the account of the Barnacle Geese giving birth to animal flesh there are also echoes of Pliny's skewed account of Theophrastus's plant-animal analogies. When Theophrastus talked about the "heart wood" or "blood" of the tree, he was clear that he was using such terms analogously in order to better understand the plant world. Pliny, however, interpreted Theophrastus's analogies between animals and plants in a literal sense.[10] Pliny's account of plant form is therefore skewed, leading to the assertion that "the bodies of the trees, as of other living things, have in them skin, blood, flesh, sinews, veins, bones and marrow."[11] Pliny also explicitly linked flowering and birth: "When a plant flowers it may be said to give birth…. This and the process of budding are the tree's labour."[12]

From the basis of Pliny's descriptions of plants with blood and flesh, and of plants giving birth, some authors thought it not too great a stretch of the imagination for plants to bear fruits containing animals made of bone and blood. In the fourteenth-century compendium of fantastical tales, *The Travels of Sir John Mandeville*, there is a story of a plant in the land of Cathay (China):

> And there groweth a manner of fruit, as though it were gourds. And when they be ripe, men cut them a-two, and men find within a little beast, in flesh, in bone, and blood, as though it were a little lamb without wool.[13]

With Pliny's plant-animal analogies in mind, I think of this account (perhaps charitably) as an act of descriptive slippage, a riffing on the status quo of plant anatomy that goes just that little bit too far! This tree bearing lambs certainly has a strong resemblance to the tale of the Barnacle Tree, something that is perhaps more than coincidental. John Mandeville, the *nom de plume* of these tales, reciprocates with a fantastical tale of his own:

> For I told them that in our country were trees that bear a fruit that become birds flying … [14]

The legend of the Barnacle Tree is an old one. John Gerard set it down in his herbal some two hundred years after Mandeville's fourteenth-century account, and the legend had been around for a long time even then, perhaps having its origins in Pliny's account of flowers giving birth. The Christian scholar Albertus Magnus expressed his doubts about such accounts of trees giving birth to birds as early as the thirteenth century.[15] Pope Pius II, in his work *Europe*, includes a folk tale of a riparian tree in Scotland whose fruits turn into ducks. The scholarly pontiff is also clearly disbelieving of this tale, wryly observing:

> When I eagerly investigated this matter, I learned that miracles always recede further into the distance and that the famous tree was to be found not in Scotland but in the Orkney islands.[16]

Regardless, numerous medieval authors have told and retold the tale of this fantastical tree, including Muenster, who speaks of a tree whose fruits are converted into living birds when falling in the water and calls this the "Goose Tree."[17] No doubt aware of some contemporary skepticism, he adds:

> Lest you should imagine that this is a fiction devised by modern writers, I may mention that all cosmographists, particularly Saxo-Grammaticus take notice of this tree.[18]

Despite such literal interpretations, the "part of truth" in these accounts is that of the shared substance and kinship connection between animals and plants explored at length in the preceding chapters. Even Pliny's literal description of the "cases of trees that talked" contains such a significant truth, that of plant sentience, which will be explored in the next chapter.

This retelling and recycling of legends is by no means confined to Europe. One of the most common legends of metamorphosis is the Arabian tale of the Wak-Wak (or waq-waq or vaq-vaq or vak-vak) tree. The story of the Wak-Wak tree appears repeatedly in the Arab legends, first appearing in a description of the mysterious island of Wak-Wak in the tenth century:

> Muhammad ibn Babishad told me that he had learned from men who had landed in the country of Waq-waq, that there is found a species of large tree, the leaves of which are round but sometimes oblong, which bear a fruit similar to a gourd, but larger and having the appearance of a human figure. When the wind shook it there came from it a human voice.... A sailor seeing one of these fruits, the form of which pleased him, cut it off to bring it back, but it immediately collapsed and there remained in his hands [only a flabby thing] like a dead cow.[19]

From these secondhand observations of a tree that bore a fruit similar to a human figure, a legend of the Wak-Wak tree arose, more mystical in character. The twelfth-century *Kitab al-Jughrafiya* (*Book of Geography*), tells of a tree in the distant islands of Waq-waq whose fruits do not simply resemble a human figure, but are human figures in the form of young girls "more beautiful than words can describe."[20]

> At the moment of falling to the ground they utter two cries: "Waq-waq!" When they have fallen to the ground, flesh without bones is found. [21]

Like any good myth or legend, details are changed whenever the tale is told. In a popular version of the *Thousand and One Nights* it says:

> After which there will be a vast mountain and a running river, which extend to the Islands of Wak Wak. On the banks of this river is a tree called Wak Wak, whose branches resemble the heads of the sons of Adam. When the sun riseth those heads all cry out: Wak Wak![22]

This noise, according to the writer Qazwini, was a portent for ill tidings: "The inhabitants of this island understand this noise and draw disagreeable omens from it."[23] Yet as with the tales of mandrake's cries, and its metamorphosis from the dead, not all were ready to take such accounts as literal descriptions. The tenth/eleventh-century author Biruni was clearly skeptical of such legends nearly six hundred years before Thomas Browne.

<blockquote>

The island of Waq-waq belongs to the Qmair islands [Qmar]. Qmair is not, as common people believe, the name of a tree which produces screaming human heads instead of fruits, but the name of a people, the colour of whom is whitish.[24]

</blockquote>

MAGICAL HEALERS

Some the world's oldest and rarest texts include descriptions of pragmatic remedies concocted from herbs and trees. In ancient Egypt, the *Ebers Papyrus* (1550 BCE) described the use of heated herbs to treat asthma. In China, the *Shennong pen Ts'ao ching* or *Great Herbal* (300 BCE–200 CE) systematically listed 365 plant remedies for a variety of physical and spiritual ailments. The *Badianus Manuscript* described a series of remedies used by the Aztecs before the Spanish conquests of the Americas, and in his *Materia Medica*, the Greek scholar Dioscorides listed approximately six hundred plants in what is perhaps the world's most influential pharmacopeia. His description of the mandrake (*Mandragora*) is characteristic of his botanical descriptions:

<blockquote>

Some boil the roots in wine until a third remains, strain it, and put it in jars. They use a winecupful of it for those who cannot sleep, or are seriously injured, and whom they wish to anaesthetise to cut or cauterize.[25]

</blockquote>

Likely building upon the work of Dioscorides, his contemporary Pliny describes the mandrake in a very similar fashion:

<blockquote>

When the mandrake is used as a sleeping draught the quantity administered should be proportioned to the strength of the patient, a moderate dose being one cyathus. It is also taken in drink for snake bite, and before surgical operations and punctures to produce anaesthesia.[26]

</blockquote>

In an analysis of the *Natural History*, E. W. Gudger contends that one of Pliny's great faults was including "in his book a great deal of hearsay data, often of improbable character."[27] In parallel with a functional pragmatism, there are instances in the *Natural History* where the descriptions of plant powers are extended into the supernatural. Pliny describes the plant "paeonia":

<blockquote>

It is said that they should be dug up by night, because to do so in the daytime is dangerous, for the woodpecker called "bird of Mars" assaults the eyes. That there is a danger, however, of prolapsus of the anus when a root is being dug up, I hold to be a very fraudulent lie, calculated to exaggerate the real facts.[28]

</blockquote>

Despite such lapses into "hearsay," Pliny largely rejects, or is at least wary of, the tales of magical herbal powers, devoting a whole book of the *Natural History* to debunk all aspects of magic. But many later texts take the functional pragmatism of earlier writings and extend the descriptions of plants back into the realm of legend. In his famous *Herbal,* Nicholas Culpeper describes the plant *moonwort,* the powers of which were debunked by Thomas Browne:

> Whatever horse casually treads upon this herb will lose his shoes; it is also said to have the virtue of unlocking their fetlocks and causing them to fall off.[29]

The plant is described with supernatural powers, but to be fair to Culpeper, he doubts the veracity of the claims. Albertus Magnus, skeptical of the tales of Geese Trees, is less circumspect in his descriptions of magical plant species. Albertus describes a plant *valeria* producing an exudate that is effective in restoring the peace between warring enemies.[30]

The mythical tree Gaokerena (also known as the Gokard Tree or White Hom Tree) is yet another plant with seemingly magical powers. In the *Bundahisn,* the Gaokerena is described in a section mysteriously described as "Conflict with Plants." In an act of creation,

> The tree of all germs was given forth, and grew up in the wide-formed ocean, from which the germs of all species of plants ever increased. And near to that tree of all germs the Gokard tree was produced, for keeping away deformed decrepitude; and the full perfection of the world arose therefrom.[31]

The Gaokerena tree is ascribed the power of keeping away decrepitude and of perfecting the world, but unlike the mandrake or the moonwort, the Gaokerena is regarded as a world tree. Rather than following Eliade's interpretation of the world tree as purely a symbol of the cosmos, in chapter 2 I opened up the possibility that these are earthlier descriptions, acknowledgments of the life and presence of real plants that help sustain life on Earth. As for the sacred plants discussed in chapter 2, the symbolism here relates not to *something else,* but to the relationships of nurture and kinship in which these plants participate.

In a later passage in the *Bundahisn,*[32] this tree is further described as a plant with an all-curing nature and with the power to grant immortality. It is considered interchangeably with *haoma* (*haoma* is the Avestan language name for the Vedic *soma*—see chapter 2), which also has such qualities. Again, the depictions are metaphorical rather than literal descriptions of plant powers. These depictions of plants with powers that extend beyond those that we are prepared to credit are clearly also entwined with the other major themes in this book including kinship and metamorphosis.

Cristoval De Molina's *Fables and Rites of the Incas* tells the story of the *lucma* tree that mysteriously makes virgin girls fall pregnant. The story involves the metamorphosis of Coniraya, the Incan god of fertility into a beautiful bird and a subsequent metamorphosis of his seed into the "likeness of a ripe and luxurious lucma."[33] A girl called Cavillaca eats the fruit and through this "she was made pregnant without other contact with man."[34]

This intimate connection between human and plant fertility is echoed by the Guambino people of southwest Colombia, who tell the story of the *yas* tree which causes women to give birth to its seeds:

> How pleasant is the perfume of the long, bell-like flowers of the Yas (*Brugmansia vulcanicola*), as one inhales it in the afternoon.... But the tree has a spirit in the form of an eagle which has been seen to come flying through the air and then to disappear.... The spirit is so evil that if a weak person stations himself at the foot of the tree, he will forget everything ... feeling up in the air as if on wings of the spirit of the Yas.... If a woman sits resting in the tree's shade, she will dream about men of the Paez tribe, and later a figure will be left in her womb which will be borne six months later in the form of pips or seeds of the tree.[35]

In contrast to the story of the *lucma* tree which causes humans to become pregnant with human offspring, contact with the Yas tree causes human beings to give birth to plants. Both depict the shared substance and genealogy of human beings and plants.

PRAYERS TO THE MEDICINAL PLANTS

Regardless of their status as legend or myth, all the stories presented in this chapter have value. When I read such stories, rather than viewing them as irrational or made up, like Thomas Browne (again, perhaps rather charitably), I see a kernel of truth that is part of a great tradition of human beings expressing their awe and gratitude toward the plant kingdom. Both legendary and mythical plants have inspired such wonder that they have been elevated in status from subservient resources, or symbols of human truths, to powerful agents akin to human beings.

In the world of plant mythology, this high esteem also inspires even more direct honoring. Although not exactly common, in our darkest hours of need, we have sometimes been known to let down our pretense of human superiority and properly acknowledge the living creatures that are keeping us alive. We don't just talk about the plants—we talk to them.

In the Anglo Saxon *Nine Herbs Charm* a variety of herbs are described as having numerous active powers, "to arrange," "to withstand," and "to fight." Amid this allegorical charm, a number of plants including mugwort, plantain, and chamomile are also addressed directly, using the personal pronoun *you*. These powerful plants are not reduced to *just* resources. Instead, their presence is acknowledged directly and they are given the respect that comes with personhood. Godfrid Storms's translation of the charm opens with a caution:

> Remember, Mugwort, what you made known,
> What you arranged at the Great proclamation.
> You were called Una, the oldest of herbs,
> you have power against three and against thirty,
> you have power against poison and against infection,
> you have power against the loathsome foe roving through the land.[36]

The charm is not just *about* nine medicinal herbs. The dramatic caution of the opening makes it clear from the outset that this charm is addressed directly *to* them. The plant in this opening, mugwort (or una) is addressed as *you*. This most useful of herbs is addressed as *thou*, in the language of Buber, or as, what I have argued elsewhere, a person. Such direct communication with healing plants also occurs in the *Kalevala* when an oak tree is questioned about the presence of honey in its branches.[37]

In India, the sacred text the *Atharva Veda* describes the use of a number of medicinal plants. The descriptions, however, are not simply of dry medicinal formulas, but contain pleas and prayers to the plants in question. A passage concerned with the sacred and medicinal hemp plant (*Cannabis sativa*) asks:

> May the hemp and may gangida protect me against vishkandha! The
> one (gangida) is brought hither from the forest, the other (hemp) from
> the sap of the furrow.[38]

The *Atharva Veda* also includes a similar prayer to the herb *kushtha* (*Saussurea costus*) to treat fever. Once more, the plant is not portrayed as a passive, mute medicinal resource. *Kushtha* is addressed as a significant, powerful being and the plea is direct:

> This person here, O kushtha, restore for me, and cure him! Render him free from sickness for me![39]

This respectful communication does not end with the gangida and kushtha plants. The *Atharva Veda* also includes a prayer (or hymn) to all the medicinal and magical plants. This prayer describes the kinship between the plants and the Earth and again beseeches the plants to use their power to cure human beings. But the prayer goes even farther, in the recognition of the power of the medicinal plants, it also recognizes their intelligence:

> Hither shall come the intelligent (plants) that understand my speech,
> that we may bring this man into safety out of misery![40]

While the idea of intelligent plants recognizing human speech may seem a step too far for most people, such an assertion is in line with the most recent scientific evidence on bioacoustics that plants can detect and react to different sounds.[41] If that is the case, then our prayers and incantations to plants are not simply a case of one-way traffic. They are the first step in a dialogue that bridges the human and the botanical worlds.

More than twenty-five years ago the philosopher Erazim Kohák explored this human dialogue with trees and its implications for our relationships with the plant kingdom. Kohák astutely recognized that if we human beings speak to trees then we engage in a

> recognition of nonhuman beings as our autonomous kin, worthy of respect. In Martin Buber's idiom, it is a matter of recognising such beings as "Thou"—not as a mute it, but a fellow being in a community of discourse.[42]

In Kohák's view, speaking to a tree is entirely appropriate as a measure of respect, for a tree is a person—that is, a being "with its own life, its own agenda, its own intrinsic worth, and worthy of respect as such."[43] The *Atharva Veda* put forward a similar position for all medicinal plants approximately two thousand years prior.

Humans may begin dialogues with plants, but we shouldn't expect them to end with some quickly uttered words, deadened into silence by the thick bark of a tree trunk. Through our utterances we open up the possibility of human-plant communication and so we should be prepared for a response to our questions and pleas. As our next chapter explores, we should also be prepared for a response that we don't necessarily want to hear, for as the anthropologist Marilyn Strathern cautions, our "relatives are always a surprise."[44]

LEGEND

THE BARNACLE TREE

There are found in the north parts of Scotland and the islands adjacent, called Orchades, certain trees, upon which grow certain shell fishes, of a white colour, tending to russet; wherein are contained little living creatures, which shells in time of maturity do open, and out of them grow those little living things; which falling into the water, do become fowls, which we call Barnacles, in the north of England brant geese and in Lancashire tree geese: but the other that do fall upon the land perish and come to nothing. Thus much by the writings of others, and also from the mouths of people of those parts, which may very well accord with truth...

There is a small island in Lancashire call the Pile of Foulders wherein are found the broken pieces of old and bruised ships, some whereof have been cast there by shipwreck, and also the trunks or bodies with the branches of old and rotten trees, cast up there likewise: whereon is found a certain spume or froth, that in time breed into certain shells, in shape like those of the muscle, but sharper pointed and of a whitish colour; wherein is contained a thing in form like lace of silk finely woven, as it were together of a whitish colour; one end whereof is fastened into the inside of the shell, even as the fish of oysters and muscles are; the other end is made fast unto the belly of a rude mass or lump, which in time comes to the shape and form of a bird... (John Gerard, *Herball*)[45]

Of the Countries and Isles that be beyond the Land of Cathay; and of the fruits there; and of twenty-two kings enclosed within the mountains

Now shall I say you, suingly, of countries and isles that be beyond the countries that I have spoken of.

Wherefore I say you, in passing by the land of Cathay toward the high Ind and toward Bacharia, men pass by a kingdom that men clepe Caldilhe, that is a full fair country.

And there groweth a manner of fruit, as though it were gourds. And when they be ripe, men cut them a-two, and men find within a little beast, in flesh, in bone, and blood, as though it were a little lamb without wool.

And men eat both the fruit and the beast. And that is a great marvel. Of that fruit I have eaten, although it were wonderful, but that I know well that God is marvellous in his works. And, natheles, I told them of as great a marvel to them, that is amongst us, and that was of the Bernakes. For I told them that in our country were trees that bear a fruit that become birds flying, and those that fell in the water live, and they that fall on the earth die anon, and they be right good to man's meat. And hereof had they as great marvel, that some of them trowed it were an impossible thing to be.

In that country be long apples of good savour, whereof be more than an hundred in a cluster, and as many in another; and they have great long leaves and large, of two foot long or more. And in that country, and in other countries thereabout, grow many trees that bear clove-gylofres and nutmegs, and great nuts of Ind, and of Canell and of many other spices. And there be vines that bear so great grapes, that a strong man should have enough to do for to bear one cluster with all the grapes. (*Travels of Sir John Mandeville*)[46]

27.

Contiguous to the above mentioned island there is an island in which a species of tree grows called vak vak. In the above mentioned island gold, from the cause of its being in great abundance, not being esteemed, the inhabitants of the island make twisted chains of it. In the island Vak Vak there are no inhabitants; sometimes with a strong wind some ships arriving, the inhabitants of the ship land on said island; where there is a species of large tree, and its constant load, and fruits fixed on its branches, are most beautiful girls, whose elegant nature and lovely persons those who see them remain astonished; and every one of them is most exactly formed like other women and hangs by the hair from the branches of the tree, like fruit. It sometimes happens that they all utter the sound vak! vak!—because of this the above-mentioned island is called Vak Vak; and whenever one of these girls becomes detached from her original place, she remains only about two days, then dying, her beautiful form is destroyed; and it is said that it sometimes happens that some men cohabit with them, and they find a most delicious odour and delightful taste. (*Tarih-i Hind-i garbi*)[47]

After which there will be a vast mountain and a running river, which extend to the Islands of Wak Wak. On the banks of this river is a tree called Wak Wak, whose branches resemble the heads of the sons of Adam. When the sun riseth those heads all cry out: "Wak Wak ! Extolled be the perfection of the King, the Excellent Creator!" In like manner also when the sun setteth those beads cry out the same words. A queen ruleth over the land and under her authority are the tribes of the Genii, Marids, and Devils, also innumerable enchanters. "Now, if thou fear, I will transport thee in a vessel, and convey thee to thine own country, but, if it be agreeable to thy heart to remain with us, I will not prevent thee." Then said Hassan: "O my mistress, I will not quit thee until I meet with my wife, or my life shall be lost." (*Arabian Nights*)[48]

القول على شجرة الواق واق

MARVELOUS MEDICINE

The hibiscum, by some persons known as the wild mallow, and by others as the "plistolochia," bears a strong resemblance to the parsnip.... The root, pulled up before sunrise, and wrapped in wool of the colour known as "native," taken from a sheep which has just dropped a ewe lamb, is employed as a bandage for scrofulous swellings ... (Pliny, *Natural History*)[49]

The moon owns this herb ... Alchemists say, that this herb is particularly useful to them in making silver. It is reported, that whatever horse casually treads upon this herb will lose his shoes; it is also said to have the virtue of unlocking their fetlocks and causing them to fall off; but whether these reports be fabulous or true, it is well-known to the country people by the name of unshoe-horse. Galen says, that, if it be given to such as are enraged by the biting of a mad dog, it does perfectly cure them. (Culpeper's, *Herbal*)[50]

LVNARIA
MINOR.
Klein Monkraut.

THE MAGIC BALSAM

Thus at last the blood-stream ended,
As the magic words were spoken.
Then the gray-beard, much rejoicing,
Sent his young son to the smithy,
There to make a healing balsam,
From the herbs of tender fibre,
From the healing plants and flowers,
From the stalks secreting honey,
From the roots, and leaves, and blossoms.

On the way he meets an oak-tree,
And the oak the son addresses:
"Hast thou honey in thy branches,
Does thy sap run full of sweetness?"
Thus the oak-tree wisely answers:
"Yea, but last night dripped the honey
Down upon my spreading branches,
And the clouds their fragrance sifted,
Sifted honey on my leaflets,
From their home within the heavens."

Then the son takes oak-wood splinters,
Takes the youngest oak-tree branches,
Gathers many healing grasses,
Gathers many herbs and flowers,
Rarest herbs that grow in Northland,
Places them within the furnace
In a kettle made of copper;
Lets them steep and boil together,
Bits of bark chipped from the oak-tree,
Many herbs of healing virtues; (*The Kalevala*)[55]

On the nature of the tree they call Gaokerena it says in revelation, that it was the first day when the tree they call Gaokerena grew in the deep mud within the wide-formed ocean; and it is necessary as a producer of the renovation of the universe, for they prepare its immortality therefrom. The evil spirit has formed therein, among those which enter as opponents, a lizard as an opponent in that deep water, so that it may injure the Haoma. And for keeping away that lizard, Ohrmazd has created there ten Kar fish which, at all times, continually circle around the Haoma, so that the head of one of those fish is continually towards the lizard. And together with the lizard those fish are spiritually fed, that is, no food is necessary for them; and till the renovation of the universe they remain in contention. There are places where that fish is written of as "the Ariz of the water"; as it says that the greatest of the creatures of Ohrmazd is that fish, and the greatest of those proceeding from the evil spirit is that lizard; with the jaws of their bodies, moreover, they snap in two whatever of the creatures of both spirits has entered between them, except that one fish which is the Vas of Panchasadvaran. This, too, is said, that those fish are so serpent-like in that deer water, they know the scratch (malishn) of a needle's point by which the water shall increase, or by which it is diminishing.

Regarding the Vas of Panchasadvaran it is declared that it moves within the wide-formed ocean, and its length is as much as what a man, while in a swift race, will walk from dawn till when the sun goes down; so much that it does not itself move the length of the whole of its great body. This, too, is said, that the creatures of the waters live also specially under its guardianship.

The tree of many seeds has grown amid the wide-formed ocean, and in its seed are all plants; some say it is the proper-curing, some the energetic-curing, some the all-curing.

Between these trees of such kinds is formed the mountain with cavities, 9999 thousand myriads in number, each myriad being ten thousand. Unto that mountain is given the protection of the waters, so that water streams forth from there, in the rivulet channels, to the land of the seven regions, as the source of all the sea-water in the land of the seven regions is from there. (*Bundahisn*)[51]

The conflict waged with plants was that when they became quite dry. Amerodad the archangel, as the vegetation was his own, pounded the plants small, and mixed them up with the water which Tishtar seized, and Tishtar made that water rain down upon the whole earth. On the whole earth plants grew up like hair upon the heads of men. Ten thousand of them grew forth of one special description, for keeping away the ten thousand species of disease which the evil spirit produced for the creatures; and from those ten thousand, the 100,000 species of plants have grown forth. From that same germ of plants the tree of all germs was given forth, and grew up in the wide-formed ocean, from which the germs of all species of plants ever increased. And near to that tree of all germs the Gokard tree [Gaokerena] was produced, for keeping away deformed (dushpad) decrepitude; and the full perfection of the world arose therefrom. (*Bundahisn*)[52]

عمل بسادن

PREGNANT BY THE LUCMA TREE

They say that in most ancient times the Coniraya Tiraco-cha appeared in the form and dress of a very poor Indian clothed in rags, insomuch that those who knew not who he was reviled him and called him a lousy wretch. They say that this was the Creator of all things; and that, by his word of command, he caused the terraces and fields to be formed on the steep sides of ravines, and the sustaining walls to rise up and support them. He also made the irrigating channels to flow, by merely hurling a hollow cane, such as we call a cane of Spain; and he went in various directions, arranging many things. His great knowledge enabled him to invent tricks and deceits touching the *huacas* and idols in the villages which he visited.

At that time they also say that there was a woman who was a *huaca*. Her name was Cavillaca, and she was a most beautiful virgin, who was much sought after by the *huacas*, or principal idols, but she would never show favour to any of them. Once she sat down to weave a mantle at the foot of a lucma tree, when the wise Coniraya succeeded in approaching her in the following manner: He turned himself into a very beautiful bird, and went up into the *lucma* tree, where he took some of his generative seed and made it into the likeness of a ripe and luxurious lucma, which he allowed to fall near the beautiful Cavillaca. She took it and ate it with much delight, and by it she was made pregnant without other contact with man. When the nine months were completed she conceived and bore a son, herself remaining a virgin; and she suckled the child at her own breast for a whole year without knowing whose it was nor how it had been engendered. At the end of the year, when the child began to crawl, Cavillaca demanded that the *huacas* and principal idols of the land should assemble, and that it should be declared whose son was the child. This news gave them all much satisfaction, and each one adorned himself in the best manner possible, combing, washing, and dressing in the richest clothes, each desiring to appear brighter and better than the rest in the eyes of the beautiful Cavillaca, that so she might select him for her spouse and husband. (*Narratives and Rites of the Incas*)[53]

Gálvez del.　　　　　　J. M. Bonifaz inc.

Achras *lucuma*.

Remember, Mugwort, what you revealed,
What you prepared at Regenmeld.
Una, you are called, eldest of herbs.
You avail against three and against thirty,
You avail against poison and against infectious sickness,
You avail against the loathsome fiend that wanders through the land.

And you, Plantain, mother of herbs,
Open from the east, mighty from within.
Over you carts creaked, over you queens rode,
Brides exclaimed over you, over you bulls gnashed their teeth.
Yet all these you withstood and fought against:

So may you poison and infectious sicknesses resist
And the loathsome fiend that wanders through the land.

Stime this herb is named; on stone it grew.
It stands against poison, it combats pain.
Fierce it is called, it fights against venom,
It expels malicious [demons], it casts out venom.
This is the herb that fought against the snake,
This avails against venom, it avails against infectious illnesses,
It avails against the loathsome fiend that wanders through the land.

Fly now, Betonica, the less from the greater,
The greater from the less, until there be a remedy for both.

Remember, Camomile, what you revealed,
What you brought about at Alorford:
That he nevermore gave up the ghost because of ills infectious,
Since Camomile into a drug for him was made.
This is the herb called Wergulu.
The seal sent this over the ocean's ridge
To heal the horror of other poison.

These nine fought against nine poisons:
A snake came sneaking, it slew a man.
Then Woden took nine thunderbolts

CHAMAEMELON
LEVCANTHEMON.
Camomill

Camillen.

C

And struck the serpent so that in nine parts it flew.
There apple destroyed the serpent's poison:
That it nevermore in house would dwell.

Thyme and Fennel, an exceeding mighty two,
These herbs the wise Lord created,
Holy in heaven, while hanging [on the cross].
He laid and placed them in the seven worlds,
As a help for the poor and the rich alike.

It stands against pain, it fights against poison,
It is potent against three and against thirty,
Against a demon's hand, and against sudden guile,
Against enchantment by vile creatures.

Now these nine herbs avail against nine accursed spirits,
Against nine poisons and against nine infectious ills,
Against the red poison, against the running poison,
Against the white poison, against the blue poison,
Against the yellow poison, against the green poison,
Against the black poison, against the blue poison,
Against the brown poison, against the scarlet poison,
Against worm-blister, against water-blister,
Against thorn-blister, against thistle-blister,
Against ice-blister, against poison-blister,

If any infection come flying from the east, or any come from the north,
Or any come from the west upon the people.
Christ stood over poison of every kind.
I alone know [the use of] running water, and the nine serpents take heed [of it].
All pastures now may spring up with herbs,
The seas, all salt water, vanish,
When I blow this poison from you.

Mugwort, plantain which is open eastward, lamb's cress, betony,
camomile, nettle, crab-apple, thyme and fennel, [and] old soap; reduce
the herbs to a powder, mix [this] with the soap and with the juice of the
apple. Make a paste of water and of ashes; take fennel, boil it in the
paste and bathe with egg-mixture, either before or after the patient applies the salve. Sing the charm on each of the herbs: three times before
he brews them, and on the apple likewise; and before he applies the
salve, sing the charm into the patient's mouth and into both his ears
and into the wound. (*Nine Herbs Charm*)[54]

Thou that art born upon the mountains, as the most potent of plants, come hither, O kushtha, destroyer of the takman, to drive out from here the takman!

To thee (that growest) upon the mountain, the brooding-place of the eagle, (and) art sprung from Himavant, they come with treasures, having heard (thy fame). For they know (thee to be) the destroyer of the takman.

The asvattha-tree is the seat of the gods in the third heaven from here. There the gods procured the kushtha, the visible manifestation of amrita (ambrosia).

A golden ship with golden tackle moved upon the heavens. There the gods procured the kushtha, the flower of amrita (ambrosia).

The paths were golden, and golden were the oars; golden were the ships, upon which they carried forth the kushtha hither (to the mountain).

This person here, O kushtha, restore for me, and cure him! Render him free from sickness for me!

Thou art born of the gods, thou art Soma's good friend. Be thou propitious to my in-breathing and my out-breathing, and to this eye of mine!

Sprung in the north from the Himavant (mountains), thou art brought to the people in the east. There the most stiperior varieties of the kushtha were apportioned.

"Superior," O kushtha, is thy name; "superior" is the name of thy father. Do thou drive out all disease, and render the takman devoid of strength!

Pain in the head, affliction in the eye, and ailment of the body, all that shall the kushtha heal-a divinely powerful (remedy), forsooth! (*Atharva Veda*)[57]

Pars 7
Curinil Lat.
മ ര ബ ൯ Mal.
कुरुवली Bram.
فحور بنغحيل Arab.

The plants that are brown, and those that are white; the red ones and the speckled ones; the sable and the black plants, all (these) do we invoke.

May they protect this man from the disease sent by the gods, the herbs whose father is the sky, whose mother is the earth, whose root is the ocean.

The waters and the heavenly plants are foremost; they have driven out from every limb thy disease, consequent upon sin.

The plants that spread forth, those that are bushy, those that have a single sheath, those that creep along, do I address; I call in thy behalf the plants that have shoots, those that have stalks, those that divide their branches, those that are derived from all the gods, the strong (plants) that furnish life to man.

With the might that is yours, ye mighty ones, with the power and strength that is yours, with that do ye, O plants, rescue this man from this disease!

I now prepare a remedy.

The plants givalâ ("quickening"), na-ghâ-rishâ ("forsooth-no-harm"), gîvanti ("living"), and the arundhatî, which removes (disease), is full of blossoms, and rich in honey, do I call to exempt him from injury.

Here come the intelligent plants that understand my speech, that we may bring this man into safety out of misery! (*Atharva Veda*)[56]

Grasses and trees
already have four phases,
namely that of sprouting out,
that of residing [and growing],
that of changing [and reproducing]
and that of dying. That is to say, this
is the way in which plants first aspire
for the goal, undergo disciplines,
reach enlightenment and enter into
extinction. We must, therefore,
regard these [plants] as belonging
to the classification of sentient beings.
Therefore when plants aspire
and discipline themselves,
sentient beings are doing so.
When sentient beings aspire
and undergo austerities,
plants are aspiring
and disciplining
themselves.
—Ryogen,
Tenth-Century
Tendai Buddhist
Teacher

5

SENTIENCE

TAKE A LOOK AT THE PLANTS AROUND YOU. Go on, have a look at that neglected house plant on the window ledge that you are always forgetting to water, or those fresh flowers on the kitchen table. If you don't have either of those, perhaps lean out of the window to catch a glimpse of that majestic tree you pass every day in the street outside. Failing that, conjure the image of your favorite tree or flower in your mind's eye.

Now, take a moment to observe these botanical forms. Look at the leaves, the stems, the petals, the bark. Now, what do you think of this flower or that tree? By that I don't mean to ask whether you like the look of it or not. I don't think there is anyone on earth who thinks plants are not aesthetically pleasing. What I mean to say is, what do you think these plants are capable of? What are their qualities and attributes? What is their potential? What can they do? Who do they do it for?

If these questions seem one part insanity to two parts heresy you probably aren't alone in your response. For most of us in "contemporary" societies, plants are just part of the scenery, a pretty stage set for the drama of human existence. The idea of plants being "capable" of anything is simply nonsensical. The very notion of plants "doing" anything is almost a joke—one that is used to good effect in apocalyptic B-movies such as *The Day of the Triffids* and *Attack of the Killer Tomatoes*.[1] Plants are passive. Plants are inert, insensitive, and mute. Plants are barely alive. They don't "do" anything! While this might just seem like common sense, there is a philosophical root to our modern views, for, of course, not all societies see plants in this way.

THIS NATURE IN A PASSIVE STATE

In *Plants as Persons*, I made the argument that in the history of Western philosophy, the view of plants as passive goes all the way back to the ancient Greek philosophers. Leaving the detailed arguments there, a brief account will suffice.

In his description of the nature of the physical world, the *Timaeus*, Plato writes of the creation of plant life from a "a nature akin to that of man" mingled "with other forms and perceptions."[2] Plato retained some elements of the mythical ancient Greek view of plant life, including the notions of metamorphosis discussed earlier and a view that plants partook of "feelings of pleasure and pain and the desires which accompany them."[3] But he goes on to discuss the very nature of the plants in question, stating:

> [T]his nature is always in a passive state, revolving in and about itself, repelling the motion from without and using its own, and accordingly is not endowed by nature with the power of observing or reflecting on its own concerns. Wherefore it lives and does not differ from a living being, but is fixed and rooted in the same spot, having no power of self-motion.[4]

This statement in the *Timaeus* greatly resembles the Genesis depiction of plant life as a passive state of existence. In *Plants as Persons*, I argue that in ancient Greece this was a major departure from plants being seen as true kin to human beings, and ruptured a cultural tradition that attributed sentience to the plant kingdom.

Plato's pupil Aristotle opened up and widened this schism even further. Aristotle did so by positing a hierarchical system for ordering life based upon three levels of soul—vegetative (or nutritive), sensitive, and rational. Plants were placed on the bottom rung below animals and humans as purely *vegetative* creatures, in the pejorative sense of the word. On the basis of their lack of similarity to human beings they were deemed to lack the "feelings of pleasure and pain" and the movement attributed to them by Plato. Plants were purely vegetative—"Plants having only the nutritive, other living beings both this and the perceptive soul."[5] As beings possessed of only the vegetative soul, plants lacked the higher-order functions of reason and intelligence, as Aristotle asserts in his work of practical ethics, *Nicomachean Ethics*:

> For the vegetative element in no way shares in reason, but the appetitive and in general the desiring element in a sense shares in it, in so far as it listens to and obeys it.[6]

Having already denied plants any attributes of perception and intelligence, Aristotle further denied them any purpose for their existence other than to be food for animals and human beings. This appropriation of plant life is aptly included in Aristotle's foundational political text, *Politics*.

In like manner we may infer that, after the birth of animals, plants exist
for their sake, and that the other animals exist for the sake of man, the
tame for use and food, the wild, if not all at least the greater part of them,
for food, and for the provision of clothing and various instruments.[7]

This philosophy has had a major (indeed, perhaps the strongest) influence
on our view of plant life. It goes something like this. Plants are passive, have no
capacity to sense or feel and have none of the attributes of intelligence. Being
"semi alive" in this way, we humans can treat them as we see fit. Certainly, the
house plant on the window ledge that hasn't been watered for weeks is outside
of our ordinary moral consideration. So much so that it's often a bit of a joke
when people admit to killing these plants. This view of plants has been buried
deeply within Western consciousness (with the help of the accounts of creation
in Genesis, and of philosophers such as Aquinas, Descartes, and Locke) and, I
would contend, is the view of plants that most of us hold today.

TRANSCENDING THE VEGETATIVE SOUL

To give us some perspective on our perception of plants, including an idea of how
unusual such a view is, we need to examine the understanding of plant capabilities in other cultural traditions. In the multitude of stories, mythologies, sayings,
philosophies, and poems in which plants feature, they are commonly recognized
as possessing many of the faculties that make human beings fully alive. In these
myths, plants transcend the purely vegetative to become perceptive, communicative, sentient creatures. This sentience is an important aspect of the recognition
of continuities between plants and human beings, and one that is now starting
to be seriously discussed by plant scientists.[8] Botanical mythologies have recognized and celebrated this sentience for thousands of years.

One way in which botanical mythologies depict this sentience is through
stories of plants demonstrating their volition, their purposive will, to pursue a course
of action. In the most unlikely of places, the Book of Judges from the Bible, there is
a striking passage with an allegorical tale about the anointment of King Abimelech.
In this tale, the different species of tree discuss their intentions to anoint a king.
Each of the species approached is reluctant to abandon its own characteristics as
a useful plant in order to take on the mantle of sovereign. Finally, the bramble
(probably not the common bramble *Rubus* spp. but more likely the tree *Ziziphus
spina-christi*, which is common in the Middle East) accepts, but warns that if the
other plants are not acting in good faith he will seek vengeance on them. While it

could be argued that such a portrayal of plants is largely symbolic, the fact that the plants in question are volitional marks this out as an interesting departure from the norm of botanical life being depicted as a passive resource.

This example of plants acting purposefully has parallels with the medieval Welsh poem *Cad Goddeu*, "the Battle of the Trees." This cryptic poem describes the time when the trees were "enchanted" by the magician Gwydion—a time in which they voiced their discomforts, raised disputes, and went into battle alongside each other. During the battle against the human beings, the trees are described as active and alert, but despite valiance and temporary victories, the battle ends in their defeat. In a line that could serve as the epitaph for humankind in a world of devastating climate change, "The mountain has become crooked, The woods have become a kiln."[9]

Many of humanity's myths involving plants relate that plants have acute perceptive and sensory abilities. The sensory awareness of trees is commonly recognized in Indigenous mythological stories, such as the Aboriginal Australian Dreaming stories explored earlier. Gunwinggu Aboriginal elder Bill Neidjie puts it best:

> Tree.
> He watching you.
> You look at tree,
> He listen to you.[10]

Such recognition is not just restricted to animistic cultures. In an Arabic culture that explicitly rejects animism, there are stories of plants that are aware of their surroundings. One of the hadith, a report of the teachings and sayings of the Prophet Mohammed, recounts the tale of a weeping date-palm tree. According to the hadith, the prophet used to stand by the date-palm on a Friday when he delivered his sermons. When followers made a pulpit for the prophet's preaching, the date-palm cried "like a pregnant she-camel," leading the prophet to climb down from the pulpit and place his hand over it.[11]

Similar depictions of perceptive and sensitive plants appear in Arab poetry. The leading ninth-century Arab poet Ibn al-Mu'tazz sought inspiration in gardens and penned these exquisite lines on the familiar *Narcissus* flowers:

> Don't you see the swaying narcissus looking at us,
> with eyes that chide us and are happy to do so,
> As if the pupils in their beauty,
> were golden vessels in camphor leaves,
> As if the dew on them were tears
> in the eyes of an abandoned lover?[12]

Such a beautiful rendering of the vitality and beauty of the *Narcissus* is matched, if not surpassed by the tenth-century poet Al-Sunawbri, who captures the presence and awareness of the garden in four lines.

Garden flowers when they smile,
they beckon ...
Still they speak though they are silent.
The silence of gardens is speech.[13]

In the ancient Greek myth of Erysichthon, the Thessalian king who chopped down the sacred grove of Demeter, the oak tree is described as having an awareness of its surroundings. In Ovid's telling of the myth in the *Metamorphoses* the oak tree is aware of being hit by the king's axe:

While he [Erysichthon] poised his axe slanting stroke, the oak of Deo [Ceres/Demeter], trembled and gave forth a groan ... [14]

These initial groans display the oak's awareness of its surroundings, but once Erysichthon begins to strike at the tree, the pouring of blood and subsequent cries reveal the oak tree's capacity to both feel and to suffer at the hands of mankind.

Such capacities for feeling and suffering are also recognized explicitly in the *Mahābhārata*. In a remarkable passage, the great Indian sage Bharadwaja, questions the sentience of trees. His skepticism is met by a lengthy reply from the sage Bhrigu who is emphatic that

[t]hey are susceptible of pleasure and pain, and grow when cut or lopped off. From these circumstances I see that trees have life. They are not inanimate.[15]

Interestingly, Bhrigu draws the conclusion that trees are alive, are animate, from the fact that they are susceptible to pleasure and plain. Equally we could say, as trees are clearly alive, they are aware of their surroundings, are capable of suffering when done harm, and of flourishing when treated well. Bhrigu also lists the sensory attributes that plants possess and draws parallels between them and the human senses of sight, sound, touch, scent, and taste.

BLACK BLOOD FROM THE BARK

In Homer's *Iliad*, the wounding (and suffering) of human beings is invariably accompanied with vivid descriptions of arterial blood issuing forth from human flesh—"Thereat shuddered the king of men, Agamennon, as he saw the black blood flowing from the wound."[16] This human possession of blood is also intimately connected with human kinship relations in the *Iliad*. In Book 6 the text states, "[T]his is the lineage and the blood whereof I avow me sprung."[17] Homer also explicitly draws parallels between human kinship and plant life.

> Even as are the generations of leaves, such are those also of men. As for
> the leaves, the wind scattereth some upon the earth, but the forest, as
> it bourgeons, putteth forth others when the season of spring is come;
> even so of men, one generation springeth up and another passeth away.[18]

For our purposes, it is interesting that blood, this most animal/human of characteristics, is linked strongly to plant life. But not only does plant mythology connect the kinship of plants with blood, but in the myths of plants that bleed when they are cut, blood is used to portray the vulnerability of plants and their capacity to suffer. This ability to bleed is a significant aspect of the plant sentience found in botanical myths.

In Virgil's *Aeneid*, a text heavily influenced by Homer, the hero Aeneas lands at Thrace, on a mission to found a colonial settlement, and there he begins to cut back the vegetation, only to find:

> For from the first tree, which is torn from the ground with broken roots,
> drops of black blood trickle and stain the earth with gore. A cold shudder
> shakes my limbs, and my chilled blood freezes with terror. Once more,
> from a second also I go on to pluck a tough shoot and probe deep the
> hidden cause; from the bark of the second also follows black blood.[19]

The myrtle tree that Aeneas wounds is the transformed human being Polydorus, son of Priam, who had been sent to Thrace on to be protected by the Thracian king. Although the tree is Aeneas in a different form, the pouring of blood from roots and bark connotes an ability to suffer on the part of the myrtle. Polydorus's botanical blood also has strong parallels with the suffering of the oak tree in the myth of Erysichthon. Once the tree is hit by the axe:

> blood came streaming forth from the,
> severed bark, even as when a huge sacrificial bull
> has fallen at the altar, and from his smitten neck
> the blood pours forth.[20]

This idea of plants that bleed/suffer is clearly connected to the ability of trees to produce sap when cut. In the Mayan sacred text the *Popol Vuh*, there is a description of the tree Chuh Cakche, also known as the dragon's blood tree (*Croton gossypifolius*) because of the bright red sap it produces when injured:

> The red sap gushing forth from the tree fell in the gourd and with it they made a ball which glistened and took the shape of a heart. The tree gave forth sap similar to blood, with the appearance of real blood. Then the blood, or that is to say the sap of the red tree, clotted, and formed a very bright coating inside the gourd, like clotted blood; meanwhile the tree glowed at the work of the maiden. It was called the "red tree of cochineal," [since then] it has taken the name of Blood Tree because its sap is called Blood.[21]

In the Islamic tradition, it is said that every year on the day that Iman Husain was martyred, a historic fir tree sheds blood in the form of tears—echoing the mournful cypress tree in the Greek myth of Cyparissus. Other plants, which produce different colored exudates, are also ascribed the ability to express their suffering through tears, a recognition once again that sentience is intertwined with metamorphosis. In the myth of Phaethon, the weeping poplar trees are the transformed Heliades, and the Myrrh tree that weeps is the princess Myrrha in a new form.

In some myths, such as that of Polydorus, the marks of sentience extend beyond blood to the pinnacle of "exceptional" human characteristics, speech. After striking the myrtle, Aeneas prays to the woodland nymphs (more about which shortly), has a think about what he's doing and then attacks the plant some more.

> [A] "piteous groan" is heard from the depth of the mound, and an answering voice comes to my ears. Woe is me! Aeneas why dost thou tear me?[22]

These protestations have reverberated for over a thousand years, appearing almost word for word in the description of trees in hell in Dante's Inferno. After encountering a wood with branches "gnarled and tangled" and with thorns of poison. Dante writes:

> Then stretched I forth my hand a little forward,
> And plucked a branchlet off from a great thorn;
> And the trunk cried, "Why dost thou mangle me?"
>
> After it had become embrowned with blood,
> It recommenced its cry: "Why dost thou rend me?
> Hast thou no spirit of pity whatsoever?"[23]

In plant mythology, speech and suffering are often intertwined—perhaps not surprisingly in view of the vulnerability of plant life at the hands of human beings. In the *Kalevala*, there is a tale of a birch tree found weeping in the forest by the hero Väinämöinen. Aware of its defenselessness against human beings, the birch tree weeps in memory of all the suffering it has had inflicted upon it. In one passage the tree recalls:

> Often unto me defenceless, Oft to me, unhappy creature,
> In the short spring come the children,
> Quickly to the spot they hurry,
> And with sharpened knives they score me,
> Draw my sap from out my body.[24]

Although the birch tree and the trees in hell are cut and ooze sap/blood, unlike the myrrh tree and the Heliades, their sentience doesn't end there. The trees talk to the human protagonist and share their woes in the hope that the human will empathize with its plight. Sadly, the birch's pleas fall on deaf ears, as Väinämöinen makes the tree into a harp. However, that does not stop the trees of the Karelian forest from expressing their suffering through speech. In a later passage when the mother of Lemminkainen is searching for her son she enquires of his whereabouts to the trees who tell her:

> We have care enough already.
> Cannot think about thy matters;
> Cruel fates have we to battle,
> Pitiful our own misfortune!
> We are felled and chopped in pieces,
> Cut in blocks for hero-fancy,
> We are burned to death as fuel,
> No one cares how much we suffer.[25]

The myth's portrayal of trees abused by humanity and left alone to suffer is deeply poignant in an age of ecosystem destruction.

TALK AMONG THE TREES

The speaking tree is perhaps the most common expression of plant sentience in the body of botanical mythology. This motif has permeated modern culture, appearing in popular literature (The Ents in Tolkien's *The Lord of the Rings* are an obvious choice), children's stories (I still empathize with

The Fir Tree in Hans Christian Andersen's tale), film and television (who can forget The Singing Bush in the movie *The Three Amigos*?), and even video games (I'm told that in the *Legend of Zelda*, the Great Deku tree is a large speaking tree).

One of the best-known instances of the talking tree is the tree that speaks with Alexander the Great on his journeys in the East. In the Persian epic poem the *Shahnama* (The Book of Kings), Alexander (Sikander) is introduced to a tree that has two trunks, one female and one male, and is told:

> One trunk will talk or ever day's ninth hour hath passed, So that the auspicious Shah will hear its voice. When it is night the female trunk will speak, its foliage will savour as 'twere musk.[26]

Enraptured with such a marvel, Alexander orders his guides to take him to the talking tree. There he is given a warning by the leaves on the bough:

> Why doth Sikandar roam so o'er the world,
> For he hath had his portion of good things,
> And, when he shall have reigned for twice seven years,
> Must quit the throne of sovereignty?[27]

Like the stories of the Barnacle Tree and the Wak-Wak Tree, this myth of Alexander and the talking tree has been retold many times. The version in the *Shahnama* is ultimately from the ancient biography of Alexander the Great, itself a text that exists in multiple versions and languages. In the *Greek Alexander Romance*, Alexander is told of a pair of trees in India "that speak with a human voice."[28] One tree is of the sun and the other is of the moon. Again, the trees issue a warning to the king:

> The span of your life is completed now, you will not be able to return to your mother, Olympias, but must die in Babylon. A short time afterwards, your mother and your wife will be murdered by your own people. Ask no more about these matters, for you will be told no more.[29]

A similar tale of a sentient tree issuing warnings is told by the Yaqui people of North America. In the Yaqui mythology, the people who lived in Yaqui country before the time of the Yaquis are known as Surems. In the time of the Surems, "One day, the people noticed a tree that seemed to be making noises in a strange language."[30] The tree spoke and its words were translated by a young girl able to understand what the tree was saying. The talking tree, an ash coloured Palo Verde (*Parkinsonia* spp.), prophesised the future and:

It warned of the coming of the white man with armour and new weapons; it told of the coming of much strife and bloodshed against these intruders and others, and of much suffering for a long time among the Surems, but that they would eventually overcome their adversaries.[31]

With such a crucial role in local culture, the talking tree is found in many retellings of this Yaqui myth.

Another large tree possessed of the power of speech appears in the *Mahābhārata*. In a description of the Garuda, a large mythical bird creature, the bird gets into a conversation with a banyan tree large (*Ficus benghalensis*), which tells it:

> Alight on this branch of mine that stretches for a hundred leagues, and eat the elephant and the tortoise.[32]

The *Mahābhārata* is also home to the myth of the *salmali* tree (*Bombax ceiba*). This text relates the story of a conversation between the *salmali* tree and the wind. Baited by the Vedic sage Narada, the *salmali* tree foolishly declares his strength and supremacy over the wind:

> My energy is more powerful and terrible than Wind's, Narada. With all his breath, Wind does not attain even an eighteenth portion of my force. When the worst Wind comes, smashing trees and mountains and whatever else there is, I stiffen against him mightily. That smasher tempest has been broken by me many times as he blew, smashing away. So, seer among Gods, I do not fear Wind even when he is angry.[33]

The wind is angered by the hubris of the *salmali* and threatens to use his force upon the tree. Overnight, the *salmali* reflects on his proclamations of strength against the wind and realizes that he is no match for the wind, nor even for some of the other trees of the forest. Fearful of what the wind will do to him, the *salmali* makes a plan, vowing to "rely upon my wits and escape Wind's threat."[34] His cunning plan of shaking off all of his leaves is, as the wind points out, exactly what the wind would have done to the *salmali*. Perhaps not the most intelligent of trees!

Many of the depictions of sentient, speaking, plants in ancient Greek mythology are connected with the existence of nymphs, briefly explored through the story of the metamorphosis of Daphne in chapter 3.

In the ancient Greek imagination, the nymphs were strongly associated with topographical features, particularly sources of water.[35] From Homer onward, the Greek myths yield a detailed taxonomy of the different types of nymphs, including naiads (nymphs of springs), oreads (mountain dwelling nymphs), hamadryads and dryads (of trees and forests).[36] Jennifer Larson notes that nymphs, "as providers of water, are naturally associated with all types of vegetation, including grasses, flowers, and above all, trees."[37] The later Greek sources mention hamadryads and dryads, but there are other tree nymphs, including the meliae (meliai), the nymphs of the mountain ash tree, which Hesiod indicated were kin with human beings.[38]

For later authors, the references to tree nymphs were to the dryads or hamadryads, as Pausanias relates in his Description of Greece, "They used to call some Nymphai (Nymphs) Dryades."[39] Jennifer Larson notes that there has been a long-held view of dryads or hamadryads as "tree spirits" that inhabit these trees.[40] As the notion of "spirit" is one side of a dualism that has "body" on the other, it is not the plant body that is sentient, but the human-like nymph spirits trapped inside the vegetative body. To me this brings to mind those literal interpreters of myth Tylor and Frazer, who dismissed animistic cultures as primitive on the basis that they recognized plants to be possessed of souls or spirits.

Taking cues from the notion of the meliai as *kin* to their tree, it is very difficult to disentangle the dryads and hamadryads and their plants in the manner of Tylor. Nymphs and plants often share the same name (e.g., Morea, the hamadryad of the mulberry tree *Morus*) and they share the same life force. Nonnus describes the nymph and the tree as "co-eval," that is, coexisting or of the same age or duration, and in his Homeric Hymns Hesiod describes the nymphs and the trees as being codependent:

> When the fate of death is near at hand, first those lovely trees wither where they stand, and the bark shrivels away about them, and the twigs fall down, and at last the life of the Nymph and of the tree leave the light of the sun together.[41]

This view of the lives of the trees and the nymphs being intertwined must have persisted for centuries, as the Libyan Greek, third-century poet Callimachus also held a similar position:

> And the earth-born Nymphe Melia wheeled about thereat and ceased from the dance and her cheek paled as she panted for her coeval oak, when she saw the locks of Helikon tremble. Goddesses mine, ye Mousai, say did the oaks come into being at the same time as the Nymphai? The Nymphai rejoice when the rain makes the oaks to grow; and again the Nymphai weep when there are no longer leaves upon the oaks.[42]

Sharing the same existence, the same life, the nymphs are clearly not separate from their plants. They are expressions of the sentience of their plants, the personified sentience to which human beings can more easily relate. When the nymphs speak and when the nymphs suffer, it is also the plants themselves that are communicating their discomfort to us. Telling the myths of these nymphs is a performance that brings this plant communication into the present day.[43]

SENTIENCE

THE TREES SEEK A KING

Listen to me, you leaders of Shechem, that God may listen to you. The trees once went out to anoint a king over them, and they said to the olive tree, "Reign over us." But the olive tree said to them, "Shall I leave my abundance, by which gods and men are honored, and go hold sway over the trees?" And the trees said to the fig tree, "You come and reign over us." But the fig tree said to them, "Shall I leave my sweetness and my good fruit and go hold sway over the trees?" And the trees said to the vine, "You come and reign over us." But the vine said to them, "Shall I leave my wine that cheers God and men and go hold sway over the trees?" Then all the trees said to the bramble, "You come and reign over us." And the bramble said to the trees, "If in good faith you are anointing me king over you, then come and take refuge in my shade, but if not, let fire come out of the bramble and devour the cedars of Lebanon." (Book of Judges)[44]

Kirſchen.
Prunus Cerasus
CCXXXIX.
L

Through language and elements:
Take the forms of the principal trees,
Arranging yourselves in battle array,
And restraining the public.
Inexperienced in battle hand to hand.
When the trees were enchanted,
In the expectation of not being trees,
The trees uttered their voices
From strings of harmony,
The disputes ceased …
The alder trees, the head of the line,
Formed the van.
The willows and quicken trees
Came late to the army.
Plum-trees, that are scarce,
Unlonged for of men.
The elaborate medlar-trees,
The objects of contention.
The prickly rose-bushes,
Against a host, of giants,
The raspberry brake did
What is better failed
For the security of life.
Privet and woodbine
And ivy on its front,
Like furze to the combat
The cherry-tree was provoked.
The birch, notwithstanding his high mind,
Was late before he was arrayed.
Not because of his cowardice,
But on account of his greatness.
The laburnum held in mind,
That your wild nature was foreign.

Pine-trees in the porch,
The chair of disputation,
By me greatly exalted,
In the presence of kings
The elm with his retinue,
Did not go aside a foot
He would fight with the centre,
And the flanks, and the rear.
Hazel-trees, it was judged,
That ample was thy mental exertion
The privet, happy his lot,
The bull of battle, the lord of the world
Morawg and Morydd
Were made prosperous in pines.
Holly, it was tinted with green,
He was the hero.
The hawthorn, surrounded by prickles,
With pain at his hand.
The aspen-wood has been topped,
It was topped in battle.
The fern that was plundered.
The broom, in the van of the army,
In the trenches he was hurt.
The gorse did not do well,
Notwithstanding let it overspread.
The heath was victorious, keeping off on all sides.
The common people were charmed,
During time proceeding of the men.
The oak, quickly moving,
Before him, tremble heaven and earth.
A valiant door-keeper against an enemy,
His name is considered.
The blue bells combined,
And caused a consternation.
In rejecting, were rejected,
Others, that were perforated.

Pear-trees, the best intruders
In the conflict of the plain.
A very wrathful wood,
The chestnut is bashful,
The opponent of happiness,
The jet has become black,
The mountain has become crooked,
The woods have become a kiln ... (*Book of Talsien*)[45]

TREE, HE WATCHING YOU

Tree same thing. E watching you.
You look at tree,
e listen to you.
E got no finger,
e can't speak

But that leaf,
e pumping, growing,
growing in the night.

While you sleeping
you dream something.
Tree and grass same thing.
They grow with your body,
with your feeling. (Bill Neidjie)[46]

ERYSICHTHON FELLS THE SACRED GROVE

This Erysichthon was a man
who scorned the gods and burnt no sacrifice on their
altars. He, so the story goes, once violated the
sacred grove of Ceres [Demeter] with the axe and profaned
those ancient trees with steel. There stood among
these a mighty oak with strength matured by centuries of growth, itself a grove. Round about it
hung woollen fillets, votive tablets, and wreaths of
flowers, witnesses of granted prayers. Often beneath
this tree dryads held their festival dances; often
with hand linked to hand inline they would encircle
the great tree whose mighty girth was full fifteen
ells. It towered as high above other trees as they
were higher than the grass that grew beneath. Yet
not for this did Triopas' son [Erysichthon] withhold his axe, as
he bade his slaves cut down the sacred oak. But
when he saw that they shrank back, the wretch
snatched an axe from one of them and said: "Though
this be not only the tree that the goddess loves, but
even the goddess herself, now shall its leafy top touch
the ground." He spoke; and while he poised his
axe for the slanting stroke, the oak of Deo [Ceres/Demeter] trembled
and gave forth a groan; at the same time its leaves
and its acorns grew pale, its long branches took on a
pallid hue. But when that impious stroke cut into
the trunk, blood came streaming forth from the
severed bark, even as when a huge sacrificial bull has
fallen at the altar, and from his smitten neck the
blood pours forth. All were astonied, and one,
bolder than the rest, tried to stop his wicked deed
and stay his cruel axe. But the Thessalian looked at
him and said: "Take that to pay you for your pious
thought!" and, turning the axe from the tree against
the man, lopped off his head. Then, as he struck

the oak blow after blow, from within the tree a voice
was heard: "I, a nymph most dear to Ceres, dwell
within this wood, and I prophesy with my dying
breath, and find my death's solace in it, that punishment is at hand for what you do." But he accomplished
his crime; and at length the tree, weakened by
countless blows and drawn down by ropes, fell and
with its weight laid low a wide stretch of woods
around.

All the dryad sisters were stupefied at their own
and their forest's loss and, mourning, clad in
black robes, they went to Ceres and prayed her to
punish Erysichthon. The beautiful goddess con-
sented, and with a nod of her head shook the fields
heavy with ripening grain. She planned in her
mind a punishment that might make men pity (but
that no man could pity him for such deeds), to rack
him with dreadful Famine. But, since the goddess
herself could not go to her (for the fates do not
permit Ceres and Famine to come together), she
summoned one of the mountain deities, a rustic
oread, and thus addressed her: "There is a place
on the farthest border of icy Scythia, a gloomy
and barren soil, a land without corn, without trees.
Sluggish Cold dwells there and Pallor, Fear, and
gaunt Famine. So, bid Famine hide herself in the
sinful stomach of that impious wretch. Let no
abundance satisfy her, and let her overcome my
utmost power to feed. And, that the vast journey
may not daunt you, take my chariot and my winged
dragons and guide them aloft." And she gave the
reins into her hands. (Ovid, *Metamorphoses*)[47]

Butea frondosa.

Bharadwaja said, "If all mobile and immobile objects be composed of these five elements, why is it that in all immobile objects those elements are not visible? Trees do not appear to have any heat. They do not seem to have any motion. They are again made up of dense particles. The five elements are not noticeable in them. Trees do not hear: they do not see; they are not capable of the perceptions of scent or taste. They have not also the perception of touch. How then can they be regarded as composed of the five (primeval) elements? It seems to me that in consequence of the absence of any liquid material in them, of any heat, of any earth, of any wind, and of any empty space, trees cannot be regarded as compounds of the five (primeval) elements."

Bhrigu said, "Without doubt, though possessed of density, trees have space within them. The putting forth of flowers and fruits is always taking place in them. They have heat within them in consequence of which leaf, bark, fruit, and flower, are seen to droop. They sicken and dry up. That shows they have perception of touch. Through sound of wind and fire and thunder, their fruits and flowers drop down. Sound is perceived through the ear. Trees have, therefore, ears and do hear. A creeper winds round a tree and goes about all its sides. A blind thing cannot find its way. For this reason it is evident that trees have vision. Then again trees recover vigour and put forth flowers in consequence of odours, good and bad, of the sacred perfume of diverse kinds of dhupas. It is plain that trees have scent. They drink water by their roots. They catch diseases of diverse kinds. Those diseases again are cured by different operations. From this it is evident that trees have perceptions of taste. As one can suck up water through a bent lotus-stalk, trees also, with the aid of the wind, drink through their roots. They are susceptible of pleasure and pain, and grow when cut or lopped off. From these circumstances I see that trees have life. They are not inanimate. Fire and wind cause the water thus sucked up to be digested. According, again, to the quantity of the water taken up, the tree advances in growth and becomes humid. In the bodies of all mobile things the five elements occur. In each the proportions are different. It is in consequence of these five elements that mobile objects can move their bodies. Skin, flesh, bones, marrow, and arteries and veins, that exist together in the body are made of earth. Energy, wrath, eyes, internal heat, and that other heat which digest the food that is taken, these five, constitute the fire that occurs in all embodied creatures. The ears, nostrils, mouth, heart, and stomach, these five, constitute the element of space that occurs in the bodies of living creatures. Phlegm, bile, sweat, fat, blood, are the five kinds of water that occur in mobile bodies. Through the breath called Prana a living creature is enabled to move. Through that called Vyana, they put forth strength for action. That called Apana moves downwards. That called Samana resides within the heart. Through that called Udana one eructates and is enabled to speak in consequence of its piercing through (the lungs, the throat, and the mouth). These are the five kinds of wind that cause an embodied creature to live and move." (*Mahābhārata*)[48]

At a distance lies the war-god's land of widespread plains, tilled by Thracians, and once ruled
by fierce Lycurgus; friendly of old to Troy, with
allied gods, as long as Fortune was ours. Hither
I pass and on the winding shore found my first city,
entering on the task with untoward fates, and from
my own name fashion the name Aeneadae.

I was offering sacrifice to my mother, Dione's
daughter, and the other gods, that they might bless
the work begun, and to the high king of the lords of
heaven was slaying a shining white bull upon the
shore. By chance, hard by there was a mound, on
whose top were cornel bushes and myrtles bristling
with crowded spear-shafts. I drew near; and essaying
to tear up the green growth from the soil, that I might
deck the altar with leafy boughs, I see an awful
portent, wondrous to tell. For from the first tree,
which is torn from the ground with broken roots, drops
of black blood trickle and stain the earth with gore.
A cold shudder shakes my limbs, and my chilled
blood freezes with terror. Once more, from a second
also I go on to pluck a tough shoot and probe deep
the hidden cause; from the bark of the second also
follows black blood. Pondering much in heart, I
prayed the woodland Nymphs, and father Gradivus,
who rules over the Getic fields, duly to bless the
vision and lighten the omen. But when with greater
effort I assail the third shafts, and with my knees
wrestle against the resisting sand—should I speak or
be silent?—a piteous groan is heard from the depth
of the mound, and an answering voice comes to my
ears. "Woe is me! why, Aeneas, dost thou tear
me? Spare me in the tomb at last; spare the

Myrtus Terentina alia.

pollution of thy pure hands ! I, born of Troy, am
no stranger to thee; not from a lifeless stock oozes
this blood. Ah! flee the cruel land, flee the greedy
shore! For I am Polydorus. Here an iron harvest
of spears covered my pierced body, and grew up
into sharp javelins." Then, indeed, with mind borne
down with perplexing dread, I was appalled, my hair
stood up, and the voice clave to my throat.

This Polydorus, with great weight of gold,
luckless Priam had once sent in secret to be reared
by the Thracian king, when he now mistrusted the
arms of Dardania and saw the city girt with siege.
When the power of Troy was crushed and Fortune
withdrew, the Thracian, following Agamemnon's
cause and triumphant arms, severs every sacred tie,
slays Polydorus, and takes the gold perforce. To
what dost thou not drive the hearts of men, O
accursed hunger for gold! When fear had fled my
soul, I lay the divine portents before the chosen chiefs
of the people, my father first, and ask what is their
judgment. All are of one mind, to quit the guilty
land, to leave a place where hospitality is profaned,
and to give our fleet the winds. So for Polydorus
we solemnize fresh funeral rites, and earth is heaped
high upon the mound; altars are set up to the dead,
made mournful with sombre fillets and black cypress;
and about them stand Ilian women, with hair streaming as custom ordains. We offer foaming bowls of
warm milk and cups of victims' blood, lay the spirit
at rest in the tomb, and with loud voice give the
last call. (Virgil, *Aeneid*)[49]

SEARCH FOR LEMMINKAINEN

Now the mother seeks her lost one,
For her son she weeps and trembles,
Like the wolf she bounds through fenlands,
Like the bear, through forest thickets,
Like the wild-boar, through the marshes,
Like the hare, along the sea-coast,
To the sea-point, like the hedgehog
Like the wild-duck swims the waters,
Casts the rubbish from her pathway,
Tramples down opposing brush-wood,
Stops at nothing in her journey
Seeks a long time for her hero,
Seeks, and seeks, and does not find him.
Now she asks the trees the question,
And the forest gives this answer:

"We have care enough already,
Cannot think about thy matters;
Cruel fates have we to battle,
Pitiful our own misfortunes!
We are felled and chopped in pieces,
Cut in blocks for hero-fancy,
We are burned to death as fuel,
No one cares how much we suffer."

Now again the mother wanders,
Seeks again her long-lost hero,
Seeks, and seeks, and does not find him.
Paths arise and come to meet her,
And she questions thus the pathways:

"Paths of hope that God has fashioned,
Have ye seen my Lemminkainen,

Has my son and golden hero
Travelled through thy many kingdoms?”
Sad, the many pathways answer:

“We ourselves have cares sufficient,
Cannot watch thy son and hero,
Wretched are the lives of pathways,
Deep indeed our own misfortunes;
We are trodden by, the red-deer,
By the wolves, and bears, and roebucks,
Driven o’er by heavy cart-wheels,
By the feet of dogs are trodden,
Trodden under foot of heroes,
Foot-paths for contending armies.” (*The Kalevala*)[50]

THE BIRCH TREE WEEPS

Väinämöinen, old and steadfast,
Then returned unto his dwelling,
Head bowed down, and sadly grieving,
And his cap awry adjusted,
And he said the words which follow:
"Unto me is lost for ever
Pleasure from the harp of pike-teeth,
From the harp I made of fish-bone."

As he wandered through the country,
On the borders of the woodlands,
There he heard a birch-tree weeping,
And a speckled tree lamenting,
And in that direction hastened,
Walking till he reached the birch-tree.

Thereupon he spoke and asked it,
"Wherefore weep'st thou, beauteous birch-tree,
Shedding tears, O green-leaved birch-tree,
By thy belt of white conspicuous?
To the war thou art not taken,
Longest not for battle-struggle."

Answer made the leaning birch-tree,
And the green-leaved tree responded:
"There is much that I could speak of,
Many things I might reflect on,
How I best might live in pleasure,
And I might rejoice for pleasure.
I am wretched in my sorrow,

And can but rejoice in trouble,
Living with my life o'erclouded,
And lamenting in my sorrow.

"And I weep my utter weakness,
And my worthlessness lament for,
I am poor, and all unaided,
Wholly wretched, void of succour,
Here in such an evil station,
On a plain among the willows.

"Perfect happiness and pleasure
Others always are expecting,
When arrives the beauteous summer,
In the warm days of the summer.
But my fate is different, wretched,
Nought but wretchedness awaits me;
And my bark is peeling from me,
Down are hewed my leafy branches.

"Often unto me defenceless
Oft to me, unhappy creature,
In the short spring come the children,
Quickly to the spot they hurry,
And with sharpened knives they score me,
Draw my sap from out my body,
And in summer wicked herdsmen,
Strip from me my white bark-girdle,
Cups and plates therefrom constructing,
Baskets too, for holding berries.

"Often unto me defenceless,
Oft to me, unhappy creature,
Come the girls beneath my branches,
Come beneath, and dance around me.
From my crown they cut the branches,
And they bind them into besoms.

"Often too, am I, defenceless,
Oft am I, unhappy creature,
Hewed away to make a clearing,
Cut to pieces into faggots.

Thrice already in this summer,
In the warm days of the summer,
Unto me have come the woodmen,
And have hewed me with their axes,
Hewed the crown from me unhappy,
And my weak life has departed.

"This has been my joy in summer,
In the warm days of the summer,
But no better was the winter,
Nor the time of snow more pleasant.

"And in former times already,
Has my face been changed by trouble,
And my head has drooped with sadness,
And my cheeks have paled with sorrow,
Thinking o'er the days of evil,
Pondering o'er the times of evil.

"And the wind brought ills upon me,
And the frost brought bitter sorrows.
Tore the wind my green cloak from me,
Frost my pretty dress from off me.
Thus am I of all the poorest,
And a most unhappy birch-tree,
Standing stripped of all my clothing,
As a naked trunk I stand here,
And in cold I shake and tremble,
And in frost I stand lamenting."

Said the aged Väinämöinen,
"Weep no more, O verdant birch-tree!
Leafy sapling, weep no longer,
Thou, equipped with whitest girdle,
For a pleasant future waits thee,
New and charming joys await thee.
Soon shalt thou with joy be weeping,
Shortly shalt thou sing for pleasure."

Then the aged Väinämöinen
Carved into a harp the birch-tree,
On a summer day he carved it,
To a kantele he shaped it,
At the end of cloudy headland,
And upon the shady island,
And the harp-frame he constructed,
From the trunk he formed new pleasure,
And the frame of toughest birchwood;
From the mottled trunk he formed it. (*The Kalevala*)[51]

"Of marvel what is there for me to note?"

The guides thus answered: "victorious Shah,
And pure of rede! there is a wonder here,
Whose match none in the world of small and great
E'er hath beheld a tree with double trunk
A marvel manifest. One trunk is female,
The other male, they speak, have boughs, and flourish.
At night the female yieldeth speech and perfume,
The male doth speak by day."
Sikandar went

With cavaliers of Rum and native chiefs

Of whom he asked: "This tree when speaketh it
Aloud?" The interpreter replied: "One trunk
Will talk or ever day's ninth hour hath passed,
So that the auspicious Shah will hear its voice.
When it is night the female trunk will speak ,
Its foliage will savour as 'twere musk."

The Shah asked: "When we pass the tree, good friend
What marvel meet we then?"

"Of going further"
The interpreter replied, "there is no question.
There is no place beyond it, and guides call it
'The World's End.'"

Then the blest Shah with his Humans
Set forward. When he reached the speaking tree
The ground was seething hot, and all its surface
Was hid by beasts' skins

"What are those, and who hath torn beasts thus?" he asked of his informant,
Who said: "The tree hath many devotees,
Who, if they hunger while engaged in worship,
Feed on the flesh of beasts."

When Sol attained
The apex of heaven's vault Sikandar heard
Aloft a cry proceeding from the leaves
Of that tall tree an awful, boding cry.
He feared and asked of the interpreter.
"Shrewd, trusty friend what say the talking leaves,
Because they bathe my heart in lymph of blood?"

The guide replied: "favourite of fortune
The leaves upon the boughs of this tree say:
'Why doth Sikandar roam so o'er the world,
For he hath had his portion of good things,
And, when he shall have reigned for twice seven years,
Must quit the throne of sovereignty?"

Sikandar wept tears of blood; the guide's words wrung his heart.
Thenceforth he spake to no one, but remained
All sorrowful till midnight. Then the loaves
Upon the other trunk gave utterance.
Again the Shah asked of his friendly guide
"What is it that the other branches say?"

Thus his informant solved the mystery:
The female branches say: "Thou travailest
In this wide world for greed and for addition.
Why torture thus thy soul? It is thy passion
To go about the world, afflicting folk

And slaying kings Thou hast not long to live,
Do not thyself cloud and contract thy days."

The great king asked of the interpreter:
"Good, discerning man inquire if I

Shall be in Eiim when my dark day shall come
And if my mother see me not alive
Will she at last enshroud this face of mine?"

The speaking tree said to the Shah: "Be speedy,
And bind the baggage on. Thy mother, kindred
In Rum, and face-veiled ladies there will look
On thee no more. Death in an alien land
Will come ere long, crown, diadem, and throne
Grow tired of thee." (*The Shahnama*)[52]

رسیدن اسکندر بر درخت گویا

Then I answered and said unto them, "If there be anything in your country which is marvellous, bring me to see it, for I desire nothing more than this, and I will deal graciously with you."

And one of their nobles said to me, "In this country we have two trees which can talk in all languages"; and I meditated within myself and said, "This is something exceedingly marvellous." And I commanded that they should beat that man with many stripes, in order that I might know the truth of his words and if they were to be believed, and when I had made him to suffer pain by the beating he said unto me, "O king, beat me not, for I do not deserve to be beaten, and stripes can make my words no better." Then straightway I believed his words, and I had compassion upon him. And I rose up with that man, and after we had marched for ten days we arrived in a place beyond which was nothing except the flower which cometh forth from Paradise; now this place is beneath the earth, and here is the abode of fiends and devils, and whosoever goeth therein they destroy. And there was none of our men who passed from us into that place that did not die. Then they went into a garden wherein there was musk, and wherein man could not enter except by the door thereof, by reason of the thickness of the trees which were therein. In the midst of that garden were figures of the sun and moon, and in it were two large hedges between which was a temple which was called "[The temple of] the sunrise and moonrise," for here the sun and the moon do rise, and inside the temple [enclosure] were two most beautiful trees, like unto which no others exist. Now it was told me that one of them was a male and the other a female; the male was the sun, and the female was the moon. Beneath them were strewn skins of lions, but I saw here no [vessel of] iron, or of stone, or of brass, or of earthenware. And again I asked them, saying, "Do both these trees speak at once, or hath each tree its own time for speech?"

And they told me that the tree of the sun speaketh at sunrise, at midday, and at sunset, each day three times, and that the tree of the moon speaketh before the hour of night, at midnight, and at the time of dawn. Then he who had charge over the sanctuary said to me, "O king, enter in to these trees," and I commanded one hundred of my friends to come in with me armed. And he who was over the sanctuary said to me, "It is not meet for thee to enter into the sanctuary with iron." Then I commanded Alexander the chief horsemen of my army to go round about outside the garden and outside the temple, and to keep guard upon me, for I was afraid lest they should plot evil against me; then I went in with fifty of my friends who were without armour.

And I cried out to one of the men who watched in the sanctuary when the tree spoke at the time of sunset, and I said to him, "Tell me [what it said], and lie not unto me, for I swear by God Almighty, that if thou deceivest me I will cut off thy head." Now he wished to hide what it said from me, but I took hold of his hand and I led him away by himself, and I sware unto him [that I would kill him]. And he drew nigh to me and called unto one of my servants, and spake unto him, "The tree saith that he is like unto a king whom one shall slay"; and I said, "God Almighty knoweth." And I went to the tree a second time at midday, and, behold, the tree spake in the Greek language that I might understand [its] words, and it said unto me, "Behold, thou shalt die in the land of Babylon," which words I heard together with my servants. Then was I struck with wonder at its words, and I was pained and grieved with an exceedingly great grief, and my heart beat wildly, and the shadow of slumber forsook my eyes during the whole night. Now when my friends and servants saw my sorrow and that I was troubled because of this thing they said unto me, "O king, if thou couldst but sleep only a very little thou wouldst be better." And in the morning I commanded my servants to come forth from that sanctuary and not to depart far away from me, and I went and drew nigh unto the [other] tree, and I said to it, "Tell me, have the days of my life drawn nigh to the end? Shall I be able to return to Macedonia, my native land, before I die?" And when the sun had risen above the tree I heard a mighty voice [speaking] in the Greek language, and saying, "Thy days are counted, and only a few of them remain unto thee. And as for thy country thou shalt nevermore reach it; thou shalt die in the land of Babylon, and thy servants shall kill thee by poison." (*The Life and Exploits of Alexander the Great*)[53]

On this they recite this ancient account of a conversation between a salmali tree [the very large *Bombax ceiba*] and the wind, O best of the Bharatas.

There was a great tree in the Snowy Mountains that was many years old and still had its trunk, its branches and its foliage. Rutting elephants would rest beneath it when plagued by the heat and worn down by exertion, and so would other kinds of animals. This tree had blossoms and fruit, and it measured a nalva [about six hundred feet] in circumference; its shade was thick, and it was filled with the noise of parrots and myna birds. Merchants in caravans and forest-dwelling ascetics and travelers would stay a while under this very charming and most excellent of trees.

O bull of the Bharatas, when Narada approached it and saw all its extensive branches and boughs, he said to it, "Oh how charming and beautiful you are! O salmali, O best of trees, you please us always. Dear, birds are always delighted to stay in your very lovely branches, and animals and elephants are delighted to stay below them. When I look at your branches and boughs and your massive trunk, I see that none of them at all have been broken by the Wind. Is Wind your cherished friend, my dear? Since surely he always protects you in this forest? Wind is powerful, and when he blows he uproots trees high and low and displaces the peaks of mountains. When Wind blows fragrant and clean he dries up even the Patala hell below the earth, and lakes and rivers and oceans. There can be no doubt, Wind must protect you because of friendship. So you have an abundance of branches and all your leaves and flowers.

"It seems very charming you, tree, that these birds are in good cheer and take their pleasure upon you. As they all sing so fetchingly in the season so favorable to flowers, I hear the sweet notes of each one of them and of all of them together. And these jolly elephants displaying the splendid beauty of their particular herds and clans have taken to you because they were plagued by the heat, and they find comfort with you, salmali. And with these other kinds of animals, and the caravans here, you shine splendidly, tree. You shine like Mount Meru! And you shine like Mount Meru because of the brahmins fully perfected through asceticism who are here. And because of the non-brahmin ascetics here too, I think this place of yours here is the equal of heaven.

"Whether it is because he is a relative or because of friendship, O salmali, there is no doubt that the terrifying Wind that sweeps everywhere watches over

Bombax heptaphyllum

you all the time. O salmali, since Wind protects you, you must have stooped to the most extreme abasement before him. You must have said to him 'I am yours.' I do not see the tree, nor the mountain firm that has not been broken by the force of the Wind. I don't think there is one on earth. But on the other hand, you and your following, salmali, are always protected by the wind for some reason. So you stand free of risk."

The salmali said:

"Wind is not my friend, brahmin, nor is he a relative, nor an ally. Nor is it that Wind is my supreme lord and thus protects me. My energy is more powerful and terrible than Wind's, Narada. With all his breath, Wind does not attain even an eighteenth portion of my force. When the worst Wind comes, smashing trees and mountains and whatever else there is, I stiffen against him mightily. That smasher tempest has been broken by me many times as he blew, smashing away. So, seer among Gods, I do not fear Wind even when he is angry."

Narada said:

"O salmali tree, your way of seeing it is just the opposite of mine, that's for sure. There is nowhere a being whose might is equal to the might of Wind, Indra, Yama, Vaisravana, and Varuna, the Lord of Waters—not even these Gods are equal to the Wind. So how could you be, tree? When any being with the breath of life in him moves upon this earth, in every instance it is the Blessed Lord Wind who causes that breath and that movement. When perfectly controlled, this one causes all the movement of living, breathing beings. On the other hand, when it is not properly controlled, it moves within men bizarrely.

"You do not respect Wind, who really is like this, who is the best of the supporters of all beings, and who is worthy of honor. What is that except lightness of mind? You have no hard inner substance, you stupid thing, you just talk a lot. O salmali, you speak vainly, cloaked in anger and such. I have become angry at you while you've been talking like this. I myself am going to tell Wind the many bad things that you've said. Stupid tree, Wind is not despised in this way by sandalwood trees, spandana trees, sala tees, pine trees, deodar trees, cane reeds, nor by tethering ropes, all of which are even stronger and more accomplished than you. These know the strength of Wind and of themselves. Therefore, these best of trees will bend to Wind's blowing. But you, from your delusion, do not know Wind's boundless power."

Bhisma said:

After that best of those who know the brahman had spoken to the salmali tree. Narada told Wind all that the salmali had said. "Some salmali tree growing on the heights of the Snowy Mountains, a tree with a great trunk and many branches and surrounded by a retinue, is contemptuous of you, Wind. He uttered many insults at you, which it would not be right for me to repeat before you, Lord Wind. Wind, I know you are the best supporter of all living beings—the very best and the most important—and I know you are like Vaivasvata in anger."

After he heard what Narada said, Wind went to the salmali tree and spoke to him angrily. "O salmali tree, the things you said before Narada expressed contempt for me. I ,Wind, will now demonstrate my power and majesty to you. I am aware of who you are. You are known to me, tree. The Lord Grandfather reached the end of his creation of beings with you. And just because he stopped with you, you have been accorded special favor. You have been protected because of that, you most stupid and lowest of trees, not because of any power of your own. Since you despise me as just another element of nature, I am going to show myself to you so you will realize who I am."

Smiling a bit, the salmali tree replied, "Wind, in your anger, go show your self to yourself out in the woods. Let go of your anger toward me. What are you going to do to me in your fury? I would not be afraid of you Wind, even if you were the Supreme Lord himself."

Wind then said, "Tomorrow I will display my energy to you."

Night then approached. The salmali tree then reflected upon things Wind had brought about and he saw that he was not equal to Wind.

"What I said to Narada about Wind was wrong. I am no match for Wind in strength: he is stronger. Wind is always mighty, as Narada said he was. In fact I am even weaker than the other trees, no doubt of that. However, no other tree is equal to me in wits, so I will rely upon my wits and escape Wind's threat. If the trees of the forest would resort to this stroke of wit they would always be safe from angry Wind, no doubt about it. Those fools do not know how Wind enraged might not blow against them, as I do."

After these reflections, the salmali tree began shaking. He knocked branches, limbs, and boughs off of himself. He saw Wind coming at dawn, after he had shed those branches and his leaves and flowers. Knocking down large trees, the angry blowing Wind came to the place where the salmali tree stood.

Wind looked at him who had no leaves, whose outer branches had fallen off, whose flowers were all gone. Smiling at him, he said to the silly salmali tree with its broken limbs:

"O salmali tree, this whole shedding of branches you have done to yourself was something I, out of anger, made you do this way. You have no flowers or outer branches, your shoots and foliage are all gone—in your own ill-conceived plan you have been in the thrall of my power."

When he heard Wind's statement the salmali was ashamed and he burned as he recalled what Narada had said to him. (*Mahābhārata*)[54]

"As for the child [Aeneas, son of Ankhises and Aphrodite], as soon as he sees the light of the sun, the deep-breasted Mountain Nymphs who inhabit this great and holy mountain shall bring him up. They rank neither with mortals nor with immortals: long indeed do they live, eating ambrosia and treading the lovely dance among the immortals, and with them the Sileni and the sharp-eyed Argus mate in the depths of pleasant caves.

"But at their birth pines or high-topped oaks spring up with them upon the fruitful earth, beautiful, flourishing trees, towering high upon the lofty mountains (and men call them holy places of the immortals, and never mortal lops them with the axe); but when the fate of death is near at hand, first those lovely trees wither where they stand, and the bark shrivels away about them, and the twigs fall down, and at last the life of the Nymph and of the tree leave the light of the sun together.

"These Nymphs shall keep my son [Aeneas] with them and rear him, and as soon as he is come to lovely boyhood, the goddesses will bring him here to you and show you your child. But, that I may tell you all that I have in mind, I will come here again towards the fifth year and bring you my son. So soon as ever you have seen him—a scion to delight the eyes—you will rejoice in beholding him; for he shall be most godlike: then bring him at once to windy Ilion. And if any mortal man ask you who got your dear son beneath her girdle, remember to tell him as I bid you: say he is the offspring of one of the flower-like Nymphs who inhabit this forest-clad hill. But if you tell all and foolishly boast that you lay with rich-crowned Aphrodite, Zeus will smite you in his anger with a smoking thunderbolt. Now I have told you all. Take heed: refrain and name me not, but have regard to the anger of the gods." When the goddess had so spoken, she soared up to windy heaven. (*Homeric Hymn to Aphrodite*)[55]

With the counsel
and advice of the latter
persons, Boniface in their presence
attempted to cut down, at a place called
Gaesmere, a certain oak of extraordinary
size called in the old tongue of the
pagans the Oak of Jupiter.
          Taking his courage in his hands
 (for a great crowd of pagans stood by watching
and bitterly cursing in their hearts the enemy of
the gods), he cut the first notch.
          But when he had made a superficial cut, suddenly,
the oak's vast bulk, shaken by a mighty blast of wind from
above crashed to the ground shivering its topmost branches
into fragments in its fall.

—Willbald, *The Life of St Boniface*

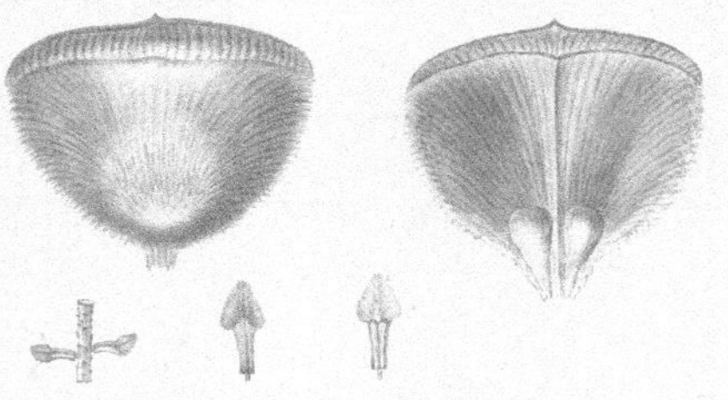

VIOLENCE

If we acknowledge that plants are sentient, that they can sense and communicate and can suffer when harmed, what does this mean for our behavior toward them? Can we continue to kill them with impunity, as St. Boniface did to the Oak of Jupiter, or should we modify our behavior accordingly?

One response to such a question might be to point to the famous is/ought gap of David Hume and argue that although plants are sentient we need not do anything to change our behavior. A factual description of plant characteristics does not mean we should behave toward them any differently. Others who may wish to avoid treating nonhumans as less than they are might be unsatisfied with such an answer. Adopting a different approach, animal rights activists largely place sentient animals as off-limits for human beings.[1] Certainly, animals acknowledged as sentient are placed within the moral sphere of "thou shalt not kill." Such a behavioral response to sentience is one that will be familiar to many people. We may not practice it personally, or necessarily agree with such a stance, but in an era of industrial meat farming, it certainly seems to be logically consistent that if you regard something as sentient, you don't eat it. We don't (as a rule) eat (sentient) humans after all!

But if plants are sentient, then what? What does that mean for how we interact with plants in the wild, in our gardens, and in our kitchens? If plants are admitted to the class of sentient beings, then killing them would also be a violent act. Most of us don't want to be continually violent in our daily lives. So, if killing sentient plants is a violent act, what on earth can we eat, then? Different religious traditions (from which we have considered myths thus far) have dealt with this issue in a number of ways.

Some Buddhist traditions, such as Theravada and Tibetan Buddhism, have found it almost impossible to reconcile plant sentience with their fundamental ethic of strict nonviolence. Instead of modifying their behavior in light of plant sentience, these traditions have instead modified plant sentience in response to human behavior. Whereas once they were recognized as active and communicative, plants have been removed from the class of "sentient beings" that are both the primary subjects of Buddhist compassion and destinations for human rebirth.

In the *Majjhima Nikāya*, a fundamental text in Theravada Buddhism, the Buddha describes the destinations of rebirth for a human being after death: "hell, the animal realm, the realm of ghosts, human beings and gods."[2] In contrast to the account of rebirth as it first appeared in the *Upaniṣads*, in Theravada plants are not considered a possible destination for rebirth, an understanding of life and death that is mirrored in the Tibetan Buddhist Bhavacakra. Denied a role in the cycle of rebirth, plants are excluded from the category of sentient beings, and the problem of sentience and violence is ignored rather than resolved.

Another approach to the problem of plant sentience can be seen in Jainism, a predominantly Indian religion that arose around the same time as Buddhism. In Jainism, in contrast to its near neighbor, plants *are* regarded as living beings. In the sacred *Acaranga Sutra* it states:

> Thoroughly knowing the earth bodies and water bodies and fire bodies and wind bodies, the lichens, seeds and sprouts, he comprehended that they are, if narrowly inspected, imbued with life.[3]

Possessed of life, plants are regarded as sentient beings, which in Jainism means that they should not be the subject of violence or killing. Yet even for Jaina clergy, for whom nonviolence toward plants is part of their strict general vow of nonviolence, there is no possibility of entirely renouncing the killing of plants. Instead of retreating from a position of plant sentience, the use of plants is acknowledged to involve killing and this is minimized wherever possible. As a result, Jain monastics are subject to some quite severe ethical restrictions in their consumption of plants (and seeds), including not accepting any raw plant foods such as sprouts, garlic, ginger, and many seeded plants such as pomegranate, as the consumption of such plants would involve direct killing.[4]

Some modern authors, such as the philosopher Michael Marder, have gone even further than the Jainas and questioned whether it is ethical to eat plants at all, if we acknowledge their sentient and intelligent natures.[5] However, before you reach for the fruitarian cookbook or pop out for a spot of photosynthesis, the canon of botanical mythology provides us with ways of exploring this thorny (no pun intended) ethical problem, without going to Marder's extremist and impractical ends.

In many botanical myths, rather than shy away from the issue, the need to kill plants is acknowledged as a fundamental aspect of human life. In Homer's *Odyssey* there is a passage in which the goddess Calypso shows the hero Odysseus a grove of trees. Once the goddess has returned homeward, "he fell to cutting timbers, and his work went forward apace. Twenty trees in all did he fell, and trimmed them with the axe."[6] Odysseus doesn't think twice about getting the timber that he needed for his vessel. In Homer, both sentience and violence are part of the relationship with the plant kingdom.

Similarly, many of the origin myths explored in chapter 1 revolve around the useful and edible plants being brought to country. In the origin myth of the Acoma, there is a long description of the use of the corn that was a staple crop of the Acoma people. Following the gift of the corn by Tsichtinako, the people are instructed how to roast it and how to use wood from the pine trees to stoke the fires. Despite their status as relations, these plants are essential for the nourishment of human beings and so are eaten.[7] As it was for Ovid, this does not involve a final ending, but simply a transformation of matter from the body of one kinsperson to another. In a world in which life and death are not binary opposites, there is no space for an insistence that one is favored over another.

In the *Atharva Veda*, there is a passage that describes the making of grain porridge for the Brahmins, the priestly caste. In this passage, the grains are poured into the pot without delay and the text is unapologetic with regard to the use of plants for such a purpose:

> Hand over the sickle, with haste bring
> promptly (the grass for the barhis); without giving
> pain let them cut the plants at the joints! They
> whose kingdom Soma rules, the plants, shall not
> harbour anger against us![8]

I find the reference in this passage to the possible anger of plants fascinating. While quite rare in the mythology of plants,[9] it also suggests an acknowledgment of plant perception and sensitivity, but also a strong autonomy, the threatening of which would possibly give rise to rage. Such recognition is also coupled with the assertion that this anger will not eventuate. The paucity of comparative material means that I can only speculate as to why this would be. A simple explanation is that the plants cannot be angry because they are clearly incapable. However, as the anger is referred to in light of the cutting being done "without giving pain," it may also be that this absence of "pain" is the reason for the absence of anger. The cutting of grass does not generally lead to the death of grass, whereas the felling of a tree often will. Another, more basic, explanation is that the human writer deems that plants cannot be angry when the taking of plant life is to feed more life—the very basis of our terrestrial ecosystems.

This flourishing of life from death is depicted in the story of Väinämöinen in *The Kalevala*. Following germination, the Oak tree of Jumala continues to grow and begins to block out the light and impede the movements of the clouds, starving and smothering the other species below. Väinämöinen approaches the tree and:

> With his axe he smote the oak-tree,
> With his sharpened blade he hewed it;
> Once he smote it, twice he smote it,
> And the third stroke wholly cleft it.[10]

The suffocating oak splinters and crashes down to earth. After the tree had fallen:

> Once again the sun shone brightly,
> And the pleasant moonlight glimmered,
> And the clouds extended widely,
> And the rainbow spanned the heavens[11]

Although Väinämöinen uses violence to take the life of the mighty Oak tree of Jumala, the myth acknowledges that in doing so a greater abundance and diversity of life is allowed to flourish. Once the sun begins to shine and the clouds begin to circulate once more:

> Then the wastes were clothed with verdure,
> And the woods grew up and flourished;
> Leaves on trees and grass in meadows.
> In the trees the birds were singing,
> Loudly sang the cheery throstle;
> In the tree-tops called the cuckoo[12]

The myth acknowledges that violence toward one species may allow many others to prosper. Another instance of flourishing following such suffocation is chronicled in a myth contained in the *Vishnu Purana*. In this story:

The trees spread and overshadowed the unprotected earth, and the people perished: the winds could not blow; the sky was shut out by the forests; and mankind was unable to labour for ten thousand years.[13]

After ten thousand years of being locked out of the forest, the sages become angry and end this smothering of life on earth—sending forth wind and flames to tear up the trees and clear them away. You might ask what took the sages so long to act, but once they did, life on earth could resume in all its abundance.

HOLDING ON TO VIOLENCE

Rather than seeking to ignore or jettison it, such myths recognize that violence toward plants is a fundamental part of our relationship with the plant kingdom. As the geographer Kathryn Yusoff has pointed out, acknowledging and embracing such imminent violence has an important role to play in our relationships with nonhumans.[14] Yusoff refers to "banal violence" as the "structural violence that is invisibly inscripted in many quotidian practices,"[15] and follows Hannah Arendt's characterization of the "banality of evil" as a "function of thoughtlessness rather than radical will."[16] This is the human thoughtlessness of making nice-smelling shampoo out of palm oil from plantations where majestic rainforest once stood. Yusoff also refers to the prior, normative violence of backgrounding the lives of nonhumans, making their lives seem "less real" and thereby diminishing the widespread systemic violence that we inflict on them. Recognizing and keeping hold of the everyday, visceral violence toward plants can lead us toward a recognition of the real capacities and capabilities of plant lives, and away from the philosophical violence of treating them as less than they are and the "banal violence" that causes large-scale levels of destruction. As Yusoff makes clear:

> Two things crucially change when violence is admitted to the frame; firstly, the risks associated with our conceptualisation of relations and the framing of subjects are brought to the fore; secondly, by shifting the visual and conceptual frame we begin to address how non-human populations are indirectly targeted and thus can begin to reconstitute our relations to eschew that banal violence.[17]

By truly acknowledging the harm that we do to plants every time we take their lives in order to eat, rather than shying away from it, we are less likely to perpetuate the unthinking domination of plants as passive resources, mere property for human use. By holding onto the violence, we are less likely to unthinkingly perpetuate the wanton destruction of plant individuals, species, and habitats. A thorough consideration and embracing of violence is then an important counterpoint to increasing our respect and care for the plant kingdom.

In our holding onto violence against plants there are a number of myths that caution against the violence required to eat well, or to shelter our families, slipping into the perpetuation of thoughtless, "banal," violence toward plants. A good example of this is found in the origin myth of the Acoma in which Tsichtinako instructs the first humans Latiku and Nautsiti how to harvest and handle the corn crop:

> Tsichtinako told them to look at it and to gather some. They saw that the corn was hard and they picked four ears. Latiku took two ears carefully without hurting the plant, but Nautsiti jerked hers off roughly. Latiku noticed this and cautioned her sister not to ruin the plants.[18]

In a similar fashion to Bill Neidjie's contemporary caution not to be too "rough" with plants, this origin myth warns against ignorantly damaging the staple corn crop by treating it carelessly, by inflicting more violence than is necessary. In this story it is striking how the human beings are told to look at the plant in question before gathering what they need. The way in which the word *look* is used introduces the idea of the humans being told to stop and consider their actions. To stop and to at least enter into the sentient presence of the plant before proceeding any frther. By doing so the Acoma people may not only reduce the violence done to the corn but also to themselves. A reduction in self-violence occurs on both pragmatic, subsistence grounds (corn being the staple crop), but also because as kin, there is a mutuality between the human and plant bodies.[19]

Polynesian mythology is replete with versions of a myth concerning Rātā, a young man who decides to build a canoe and goes into the forest to cut a tree to work with.[20] John White's *Ancient History of the Māori* contains a version of the myth of Rātā, the son of Wahieroa.[21] Following the killing of his father by crane (Matuku), Rātā looks to make a canoe to pursue his father's killer. Rātā's mother advises her son to take his axe to the grindstone and then to go in search of a grand tōtara tree (*Podocarpus totara*) in the forest. The myth tells of Rātā cutting down the tree, but doing so in ignorance of the tree's status as revered ancestor, Tāne, the God of the Forest. The tree magically returns to an upright state "as though it had never been cut down."[22] Rātā goes once more into the forest and cuts down the tree, only to be told by "some being":

> You ignorantly cut down
> The sacred forest of Tāne,
> The sacred chips of Tāne.[23]

The gods are angry with Rātā, not for cutting down the tree to use as a canoe, but for doing so ignorantly, without acknowledging the violence that he was perpetuating on the forest of his ancestor Tāne and without offering the correct propitiation—the business of "seeking permission and offering placation," which Māori still do to this day.[24] The consequences for Rātā are not particularly harsh, but the myth acts as a strong reminder of the presence of the tōtara and the need to acknowledge the taking of life.

In a survey of the Rātā myth across Polynesia, Kozmin contrasts the existence of versions in which the tree is re-erected, with one from Rarotonga in which the tree is not brought back:

> Rata went to the forest and commenced to hollow out the canoe, and when night came on he hid under the branches he had cut off. He had not been long in his hiding-place when the gods, Atonga and Tongaiti-matarau, appeared, and called upon the tree to resume its natural position and state; these are the words they used:—
>
> "Piri mai, piri mai taku māieti, taku māieta,—
> Tū, tū te rau tu."
> "Join together, come together,
> My beloved, my cherished ones—
> Rejoin your parents O leaves."
> The fallen tree did not respond to the command, so again these gods called:—
> "Piri mai, piri mai taku māieti, taku māieta;—
> Tū, tū te rau tu."
> "Join together, come together,
> My beloved, my cherished ones—
> Rejoin your parents O leaves."
>
> Still the tree did not respond, and when Atonga and Tongaiti-matarau looked about to discover the cause why the tree would not obey their command, they saw Rata hidden under the branches; they could see his eyes glistening in the darkness.[25]

This is a stark reminder of the need for propitiation. The consequences of Rātā's ignorance are that the tree is killed, both materially and spiritually, both cut down and cut off from the powers of the Gods and their ability to affect the bodies of living beings.

CONSEQUENCE OF SENSELESS VIOLENCE

In direct contrast to those myths in which thoughtless violence is frowned upon, there are myths in which violence toward plants is committed senselessly. In the *Mahābhārata*, some of the most striking passages of violence toward plants involve the Pandava brothers, in particularly the fights between the second brother Bhima and the *rakshasas* (demons) during the brothers' exile in the forest. In these fights, Bhima uses his herculean strength to rip up the trees and use them as weapons, destroying whole tracts of forest in these epic battles. In one gargantuan fight

> the trees struck at the heads of the combatants, were broken into shivers, like lotus-stalks thrown on the temples of infuriate elephants. And in that great forest, innumerable trees, crushed like unto reeds, lay scattered as rags.[26]

There is a resemblance here to a passage in the *Epic of Gilgamesh* in which the protagonist and his companions attack Humbaba (Huwawa), the guardian of the cedar forest in which the gods live. After slaying Humbaba, Gilgamesh and Enkidu lay waste to the cedar forest. They purposefully fell the forest in what, even thousands of years since the narrative was created, resonates as a clearly violent assault on nature. With the guardian of the forest gone, the trees themselves are laid low, turned from majestic beings to *timber* for human use in mere minutes. This ancient epic sadly contains a very familiar tale of our forests' vulnerability to human violence.

Where Gilgamesh hints, many myths explain the severe consequences for inflicting ignorant violence on plant life. One of the most interesting mythological accounts concerning the consequences of conducting banal, senseless violence on plants comes from the Ancient Egyptian *Tale of Two Brothers*. The passage of interest begins with the account of botanical splendors in the kingdom:

> "Two great Persea trees have grown, as a great marvel of his majesty, in the night by the side of the great gate of his majesty." And there was rejoicing for them in all the land, and there were offerings made to them.[27]

The King and the Princess go out on horseback to marvel at the tall *persea* trees (the NE African/Arabian tree *Mimusops laurifolia*) and as they sit beneath the trees, one of them begins to speak:

I am Bata, I am alive, though I have been evilly entreated. I knew who caused the acacia to be cut down by Pharaoh at my dwelling. I then became an ox, and thou caused that I should be killed.[28]

Many days later, the Princess, nursing some inner violence, asks the King to cut down the *persea* tree. The King acquiesces to her wishes and he sends out a team of craftsmen who "did all that was in her heart unto the trees."[29] Such wanton destruction was not without its consequences though. As the workmen cut at the tree, a chip of wood flies up and the Princess swallows it. Echoing the earlier myth of the *lucma* tree, this wood causes the Princess to become pregnant with a son. Despite her lack of care and the wanton destruction of the trees, the Princess cannot escape her kinship relationship with the plant kingdom. In the stead of the *persea* trees she is forced to care for a son who is their direct offspring.

Other unusual consequences feature in the Japanese epic story, the *Nihongi*. The official Kahabe no Omi is sent into the mountains with an order to build ships for the emperor. After finding a good supply of timber he is about to cut down a fine specimen when a man warns him that the tree is a tree sacred to the thunder god: "This is a thunder-tree, and must not be cut."[30] Kahabe no Omi questions the right of the thunder god to oppose the emperor's commands but, no longer ignorant of the tree's nature, he offers "many mitegura"—votive offerings.[31] Still, when his workmen proceed to cut down the tree, the god shows his displeasure by sending down a thunderstorm.

For cultures in which kinship manifests as a "mutuality of being," the consequences of senseless violence toward the plant are writ large on the human body. Big Bill Neidjie makes clear that the shared substance of humans and plants means that our bodies cannot escape the physical consequences of our actions:

That tree for us, for anybody
but some places they don't like it tree.
They say…

"Oh that tree can cut down.
Throw im out."
No.

Because that tree e cutting … that his feeling
If we chop im that tree,
    might be bad headache
    or feel funny his body
because e cutting it our body.[32]

The consequences of cutting some trees are even more severe. In the Northern Territory of Australia, the *mabara* trees that are found at Mandibin are the transformed sons of Mananda who sat down in Gunwinggu country and turned into a yam.[33] They are *djang* (taboo), and no one may touch these trees. In the same story, some of the Waranoidjagu sons went on to Mangaranggi and died after falling into a heavy sleep. The brothers turned into a "hill with jungle trees" and once again, these trees may not be touched.

No one can touch that jungle or the trees in it, because these are the people who died. Anyone who touches them will die too.[34]

The *Vishnu Purana* cautions against the wanton, scorched earth destruction of the type found in the *Mahābhārata*. Following the clearing of the forest by the sages, the "sovereign of the vegetable world," the sacred plant soma restrains the patriarchs from destroying the few remaining trees. Instead, soma calms the men and forms "an alliance" between them and the trees through Marisha, the "daughter of the woods."[35] Instead of the self-harm of total domination over the plant kingdom, this explicit kinship alliance of human-plant helps both the plants and humanity prosper. By restraining our violence toward plants the myth acknowledges that we allow all life on earth, including human life, to flourish.

Human beings are totally dependent on the plant kingdom for food, energy, and shelter. Our lives are such that we do not have the choice of eschewing the use of plants. The severe dietary restrictions of the Jaina are not a path that many on earth can take. The only way to completely "avoid" violence toward plants is to go on pretending that they are not sentient, aware, and intelligent—a more complete denial than the "periodic strategic amnesia" described by Harvey in some animist cultures.[36] But, as we have seen, this is the path of life-denying, philosophical violence, which smooths the way for the banal, unthinking, scorched-earth policy that we have inflicted on our plants and our planet.[37]

If we acknowledge the sentience of the plant kingdom (an idea that is now being seriously mooted in the plant sciences),[38] and we accept the violence that we take part in, we can be prompted to seek ways to reduce it. Holding both sentience and violence in mind whenever we approach a plant to use its leaves, roots, or shoots, we can restrain ourselves, taking only what we need and not killing wantonly. If we hold this violence close when preparing and eating our food, we may be prompted to eat only what we need and to minimize waste wherever possible. One final time, mythology provides a sharp and succinct rendering of such a way of engaging with plant life. In the *Nihongi*, when the Prince Imperial sees death he spontaneously relates the essence and effect of holding onto violence toward plants:

> In my heart I thought
> To cut thee,
> In my heart I thought
> To take thee …
> Uncut I leave thee,
> O thou Mayumi tree[39]

VIOLENCE

DESTROYING PLANTS

See! there are men who control themselves; others pretend only to be houseless, for one destroys this (body of a plant) by bad and injurious doings, and many other beings, besides, which he hurts by means of plants, through his doing acts relating to plants. About this the Revered One has taught the truth: for the sake of the splendour, honour, and glory of this life, for the sake of birth, death, and final liberation, for the removal of pain, man acts sinfully towards plants, or causes others to act so, or allows others to act so. This deprives him of happiness and perfect wisdom. About this he is informed when he has understood, or heard from the Revered One or from the monks, the faith to be coveted. There are some who, of a truth, know this (i.e., injuring) to be the bondage, the delusion, the death, the hell. For this a man is longing when he destroys this (body of a plant) by bad and injurious doings, and many other beings, besides, which he hurts by means of plants, through his doing acts relating to plants. Thus I say.

As the nature of this (i.e., men) is to be born and to grow old, so is the nature of that (i.e., plants) to be born and to grow old; as this has reason, so that has reason; as this falls sick when cut, so that falls sick when cut; as this needs food, so that needs food; as this will decay, so that will decay; as this is not eternal, so that is not eternal; as this takes increment, so that takes increment; as this is changing, so that is changing. He who injures these (plants) does not comprehend and renounce the sinful acts; he who does not injure these, comprehends and renounces the sinful acts. Knowing them, a wise man should not act sinfully towards plants, nor cause others to act so, nor allow others to act so. He who knows these causes of sin relating to plants, is called a reward-knowing sage. Thus I say. (*Acaranga Sutra*)[40]

ODYSSEUS'S RAFT

As soon as early Dawn appeared, the rosy-fingered,
straightway Odysseus put on a cloak and a tunic,
and the nymph clothed herself in a long white robe,
finely woven and beautiful, and about her waist she
cast a fair girdle of gold, and on her head a veil
above. Then she set herself to plan the sending of
the great-hearted Odysseus. She gave him a great
axe, well fitted to his hands, an axe of bronze,
sharpened on both sides; and in it was a beautiful
handle of olive wood, securely fastened; and there-
after she gave him a polished adze. Then she led
the way to the borders of the island where tall
trees were standing, alder and poplar and fir, reaching to the skies, long dry and well-seasoned, which
would float for him lightly. But when she had
shewn him where the tall trees grew, Calypso, the
beautiful goddess, returned homewards, but he fell
to cutting timbers, and his work went forward apace.
Twenty trees in all did he fell, and trimmed them
with the axe; then he cunningly smoothed them all
and made them straight to the line. Meanwhile
Calypso, the beautiful goddess, brought him augers;
and he bored all the pieces and fitted them to one
another, and with pegs and morticings did he hammer
it together. Wide as a man well-skilled in carpentry
marks out the curve of the hull of a freight-ship,
broad of beam, even so wide did Odysseus make his
raft. And he set up the deck-beams, bolting them
to the close set ribs, and laboured on; and he
finished the raft with long gunwales. In it he set a
mast and a yard-arm, fitted to it, and furthermore
made him a steering-oar, wherewith to steer. Then he
fenced in the whole from stem to stern with willow
withes to be a defence against the wave, and strewed
much brush thereon. (Homer, *Odyssey*)[41]

ZEA MAÏS (Varietates)

After some time the corn ripened. Tsichtinako told them to look at it and to gather some. They saw that the corn was hard and they picked four ears. Latiku took two ears carefully without hurting the plant, but Nautsiti jerked hers off roughly. Latiku noticed this and cautioned her sister not to ruin the plants. They took the ears of corn to Tsichtinako saying, "We have brought the corn, it is ripe." Tsichtinako agreed and explained that the corn ears when cooked would be their food. They did not understand this and asked what they would cook with. Tsichtinako then told them that Uchtsiti would give them fire. That night as they sat around they saw a red light drop from the sky. After they had seen it, Tsichtinako told them it was fire, and that they were to go over and get some of it. They asked with what, and she told them to get it with a flat rock because it was very hot and they could not take it in their hands. After getting it with a rock, they asked what they were to do with it, and were told they were to make a fire, to go to the pine tree they had planted, to break off some of the branches and put them in the fire. They went to the tree and broke some of the twigs from it. When they got back to the fire, they were told to throw the twigs down. They did so and a large pile of wood appeared there. Tsichtinako told them this wood would last many years till there was time for trees to grow, and showed them how to build a fire. She told them that with the flames from the fire they would keep warm and would cook their food.

Tsichtinako next taught them how to roast the corn. "When it is cooked," she explained, "you are to eat it. This will be the first time you have eaten, for you have been fasting for a long time and Uchtsiti has been nourishing you. You will find salt in your baskets; with this you will season the corn." They began to look for this and Tsichtinako pointed it out to them. As soon as they were told this, Nautsiti grabbed some corn and salt. She was the first to taste them and exclaimed that they were very good, but Latiku was slower. After Nautsiti had eaten part, she gave it to Latiku to taste. When both had eaten, Tsichtinako told them that this was the way they were going to live and be nourished. They were very thankful, saying, "You have treated us well." They asked if this would be their only food. Tsichtinako said, "No, you have many other things in your baskets; many seeds and images of animals, all in pairs. Some will be eaten and taken for nourishment by you." After they had used the salt, they were asked by Tsichtinako to give life to this salt by praying to the Earth, first in

the North direction, then in the West, then in the South, and then in the East,
And when they did so, salt appeared in each of these directions. Tsichtinako
then instructed them to take always the husks from the corn carefully and to
dry them. They were then instructed to plant *ha'mi* (tobacco). When the plant
matured, they were taught how to roll the leaves in corn husks and to smoke
it. (Orgin Myth of the Acoma)[42]

STIR UP THE GRAINS

Whether ye are over-abundant or just sufficient, ye are surely clear, pure, and
    immortal: cook, ye waters, instructed by the husband and wife,
obliging and helpful, the porridge!

Counted drops penetrate into the earth,
commensurate with the breaths of life and the plants.
The uncounted golden (drops), that are poured into
(the porridge), have, (themselves) pure, established
complete purity.

The boiling waters rise and sputter, cast up
foam and many bubbles. Unite, ye waters, with
this grain, as a woman who beholds her husband in
the proper season!

Stir up (the grains) as they settle at the
bottom: let them mingle their inmost parts with
the waters! The water here I have measured with
cups; measured was the grain, so as to be according
to these regulations.

Hand over the sickle, with haste bring
promptly (the grass for the barhis); without giving
pain let them cut the plants at the joints! They
whose kingdom Soma rules, the plants, shall not
harbour anger against us!

Strew a new barhis for the porridge: pleasing
to its heart, and lovely to its sight it shall be! Upon
it the gods together with the goddesses shall enter;
settle down to this (porridge) in proper order, and
eat it! (*Atharva Veda*)[43]

For the boat of Väinämöinen,
And a keel to suit the minstrel?
Pellervoinen, earth-begotten,
Sampsa, youth of smallest stature,
He shall seek for timber for him,
And shall seek an oak-tree for him,
For the boat of Väinämöinen,
And a keel to suit the minstrel.

So upon his path he wandered
Through the regions to the north-east,
Through one district, then another,
Journeyed after through a third one,
With his gold axe on his shoulder,
With his axe, with copper handle,
Till he found an aspen standing,
Which in height three fathoms measured.

So he went to fell the aspen,
With his axe the tree to sever,
And the aspen spoke and asked him,
With its tongue it spoke in thiswise:

"What, O man, desire you from me?
Tell your need, as far as may be."

Youthful Sampsa Pellervoinen,
Answered in the words which follow:
"This is what I wish for from thee,
This I need, and this require I,
'Tis a boat for Väinämöinen;
For the minstrel's boat the timber."

And the aspen said astounded,
Answered with its hundred branches:

"As a boat I should be leaking,
And would only sink beneath you,
For my branches they are hollow.
Thrice already in this summer,
Has a grub my heart devoured,
In my roots a worm has nestled."

Youthful Sampsa Pellervoinen
Wandered further on his journey,
And he wandered, deeply pondering,
In the region to the northward.

There he found a pine-tree standing,
And its height was full six fathoms,
And he struck it with his hatchet,
On the trunk with axe-blade smote it,
And he spoke the words which follow:
"O thou pine-tree, shall I take thee,
For the boat of Väinämöinen,
And as boatwood for the minstrel?"

But the pine-tree answered quickly,
And it cried in answer loudly,

"For a boat you cannot use me,
Nor a six-ribbed boat can fashion,
Full of knots you'll find the pine-tree.
Thrice already in this summer,
In my summit croaked a raven,
Croaked a crow among my branches."

Youthful Sampsa Pellervoinen
Further yet pursued his journey,
And he wandered, deeply pondering,
In the region to the southward,

Till he found an oak-tree standing,
Fathoms nine its boughs extended.

And he thus addressed and asked it:
"O thou oak-tree, shall I take thee,
For the keel to make a vessel,
The foundation of a warship?"

And the oak-tree answered wisely,
Answered thus the acorn-bearer:
"Yes, indeed, my wood is suited
For the keel to make a vessel,

Neither slender 'tis, nor knotted,
Nor within its substance hollow.
Thrice already in this summer,
In the brightest days of summer,
Through my midst the sunbeams wandered.
On my crown the moon was shining,
In my branches cried the cuckoos,
In my boughs the birds were resting."

Youthful Sampsa Pellervoinen
Took the axe from off his shoulder,
With his axe he smote the tree-trunk,
With the blade he smote the oak-tree,
Speedily he felled the oak-tree,
And the beauteous tree had fallen.

First he hewed it through the summit,
All the trunk he cleft in pieces,
After this the keel he fashioned,
Planks so many none could count them,
For the vessel of the minstrel,
For the boat of Väinämöinen.

Then the aged Väinämöinen,
He the great primeval sorcerer,
Fashioned then the boat with wisdom,
Built with magic songs the vessel,
From the fragments of an oak-tree,
Fragments of the shattered oak-tree. (*The Kalevala*)[44]

Whilst the Prachetasas were thus absorbed in their devotions, the trees spread and overshadowed the unprotected earth, and the people perished: the winds could not blow; the sky was shut out by the forests; and mankind was unable to labour for ten thousand years. When the sages, coming forth from the deep, beheld this, they were angry, and, being incensed, wind and flame issued from their mouths. The strong wind tore up the trees by their roots, and left them sear and dry, and the fierce fire consumed them, and the forests were cleared away. When Soma (the moon), the sovereign of the vegetable world, beheld all except a few of the trees destroyed, he went to the patriarchs, the Prachetasas, and said, "Restrain your indignation, princes, and listen to me. I will form an alliance between you and the trees. Prescient of futurity, I have nourished with my rays this precious maiden, the daughter of the woods. She is called Marisha, and is assuredly the offspring of the trees. She shall be your bride, and the multiplier of the race of Dhruva. From a portion of your lustre and a portion of mine, oh mighty sages, the patriarch Daksha shall be born of her, who, endowed with a part of me, and composed of your vigour, shall be as resplendent as fire, and shall multiply the human race." (*Vishnu Purana*)[45]

RĀTĀ

Matoka-rau-tawhiri went to collect firewood, and sought
and found a tree—a beautiful tree, a grand totara tree—
some twigs of which she brought in her hand to the settlement, and when evening came she spoke to her son Rata
and said, "I have seen a fine tree—a totara-tree: on the
morrow you must go and see it." And she gave him the
twigs she had brought from the tree. He went, but could
not find it, and came back to his mother and said, "I cannot see the tree you speak of." She said, "You cannot
mistake it: it is the rough-barked tree which you will see."
Again he went, and came back; but in the third attempt he
found it, and asked his mother, "What action shall I
take?" She gave him some stone axes, but he complained
of their being blunt, and without teeth. His mother said,
"Go and hold the axe upon the back of your ancestor who
is called Hine-tu-a-oaka (Hine-tu-a-hoanga—daughter of
the whetstone), who, when you put it on her, will say, 'Be
sharpened, be sharpened, be sharpened,' and your axe will
become sharp; then you can take it to your house and
put a handle on it." He slept, and at dawn of day he
went and felled the tree, and cut off the top of it, and
came back and stayed in his house. On the morrow he
returned to the tree, which he found had been put up
again as though it had never been cut down. He again
cut it down and cut the top off, and came back to his
house, and said to his mother, "When I went to the tree,
it was standing as though it had never been cut down."

She asked, "What did you do to the tree?" He said, "I
cut it down without having first propitiated the gods by
performing the usual ceremonies and repeating the incantations
for such an act." She said, "It is not well to cut
your ancestor down ignorantly." He said, "Yes, I at
once cut it down, without any ceremony." She said,

Metrosideros robusta, A. Cunn.

"Go, return." He went and cut it down again, and cut
the top off, and went on one side and hid himself, and
heard these words repeated by some beings:—

It is Rata—it is Rata of Wahie-roa.

You ignorantly cut down

The sacred forest of Tane,

The sacred chips of Tane.

The chips of the root fly,

The chips of the top fly.

They adhere, they go near.

They are all bound on again.

Stand up and wave (0 tree! in the wind).

The tree again stood in its old position. He rushed out
and stopped those creatures, who flew hither and thither on
every side of him, and left the tree. He said, "Why do
you meddle with my tree?" They replied, "Go, return to
your home; leave your tree here, and we will make your
canoe." He went home, and his mother asked, "What is
the state of your tree?" He answered, "I found it standing up again, and cut it down, and cut off its head, and
stood aside and watched, and heard my name repeated."
He slept, and on awakening on the morrow found a canoe
had been made and brought to the side of his house. (John White, *The Ancient
History of the Māori*)[46]

Filled with wrath at this, the Rakshasa[47] struck, from behind with both his arms a heavy blow on the back of Vrikodara, the son of Kunti. But Bhima, though struck heavily by the mighty Rakshasa, with both his hands, did not even look up at the Rakshasa but continued to eat as before. Then the mighty Rakshasa, inflamed with wrath, tore up a tree and ran at Bhima for striking him again. Meanwhile the mighty Bhima, that bull among men had leisurely eaten up the whole of that food and washing himself stood cheerfully for fight. Then, O Bharata, possessed of great energy, Bhima, smiling in derision, caught with his left hand the tree hurled at him by the Rakshasa in wrath. Then that mighty Rakshasa, tearing up many more trees, hurled them at Bhima, and the Pandava also hurled as many at the Rakshasa. Then, O king, the combat with trees between that human being and the Rakshasa, became so terrible that the region around soon became destitute of trees. Then the Rakshasa, saying that he was none else than Vaka, sprang upon the Pandava and seized the mighty Bhima with his arms.

That mighty hero also clasping with his own strong arms the strong-armed Rakshasa, and exerting himself actively, began to drag him violently.

Dragged by Bhima and dragging Bhima also, the cannibal was overcome with great fatigue. The earth began to tremble in consequence of the strength they both exerted, and large trees that stood there broke in pieces. Then Bhima, beholding the cannibal overcome with fatigue, pressed him down on the earth with his knees and began to strike him with great force. Then placing one knee on the middle of the Rakshasa's back, Bhima seized his neck with his right hand and the cloth on his waist with his left, and bent him double with great force. The cannibal then roared frightfully.

And, O monarch, he also began to vomit blood while he was being thus broken on Bhima's knee. (*Mahābhārata*)[48]

Vidura continued, "Thus addressed by the Rakshasa, the virtuous Yudhishthira, steadfast in his pledges, said, 'It can never be so,—and in anger rebuked the Rakshasa.' The mighty-armed Bhima then tore up in haste a tree of the length of ten Vyasas and stripped it of its leaves.

And in the space of a moment the ever-victorious Arjuna stringed his bow Gandiva possessing the force of the thunderbolt. And, O Bharata, making

Jishnu desist, Bhima approached that Rakshasa still roaring like the clouds and said unto him, 'Stay! Stay!' And thus addressing the cannibal, and tightening the cloth around his waist, and rubbing his palms, and biting his nether lip with his teeth, and armed with the tree, the powerful Bhima rushed towards the foe. And like unto Maghavat hurling his thunderbolt, Bhima made that tree, resembling the mace of Yama himself descend with force on the head of the cannibal. The Rakshasa, however, was seen to remain unmoved at that blow, and wavered not in the conflict.

"On the other hand, he hurled his lighted brand, flaming like lightning, at Bhima. But that foremost of warriors turned it off with his left foot in such a way that it went back towards the Rakshasa. Then the fierce Kirmira on his part, all on a sudden uprooting a tree darted to the encounter like unto the mace bearing Yama himself. And that fight, so destructive of the trees, looked like the encounter in days of yore between the brothers Vali and Sugriva for the possession of the same woman. And the trees struck at the heads of the combatants, were broken into shivers, like lotus-stalks thrown on the temples of infuriate elephants. And in that great forest, innumerable trees, crushed like unto reeds, lay scattered as rags. That encounter with trees between that foremost of Rakshasas and that best of men, O thou bull of the Bharata race, lasted but for a moment." (*Mahābhārata*)[49]

Mananda had two brothers, Bidjilig (salt-water Bream) and Dadbi (Brown snake). Although not mentioned before, they came with Mananda and his Waranoidjagu sons. The sons said "Father, wait here till we return." With their father's two brothers, they went on to Magalg. They stayed there for a while, but then went on making a plain all the way to Ma_aidj-balwal (salt pan) and eastward to Gw'rela [Maung]. One of the sons had a bad cold, so they left him there. He just lay down and died, and turned into a hill. [If anyone touches that hill, he or she will get a bad cold.] The others went on to Abudjareid, where they stayed for a short time, and then on to Udjurulg. Here they left another son. He too got a bad cold and died, and turned into cough Dreaming. [There is a hill at this place, "like a djang. You can't touch it."]

Leaving that place they went on to Ilwaidu, then on to Guri-ngalwilgbil and on to Mauwudjimi [still in Maung territory]. No longer following the river, they went into bush country, eventually reaching Mandibin. Four of the sons became sick here: they "turned to" cough djang, becoming *mabara* trees, similar to paperbark. [No one may touch these.]

The remaining men went on to Mangaranggi. "It is better for us to stay here," they agreed. They made a ceremony. The Waranoidjagu sons were all song-men, and they danced while their father's two brothers sang the Waranoidjagu cycle. A large group of people from the east, the south and the west came into Mangaranggi for this big ceremony. All those people were camping there. Everyone became sick. Only the father's two brothers were all right. They all prepared to rest. But they fell into a heavy sleep, and as they slept they died. All of them "turned into" a hill with jungle trees.

The father's two brothers remained alive. That place is called "cough." No one can touch that jungle or the trees in it, because these are the people who died. Anyone who touches them will die too. The name of that place is Mangaral-rindjii [in both Maung and Gunwinggu territories]. And the father's two brothers spoke, "No-one can cut these trees—this is a danger place!" They too stayed here. "They were human; they were frightened. They remain there alive, as spirits." (Gunwinggu Dreaming)[53]

CEDRVS *foliis rigidis acuminatis non deciduis, conis subrotundis erectis* Plant. sel. tab. 1.

a. *juli gemma; b. julus cum calyce pronascens; c. idem per longitudinis medium dissectus, d. in magnitudine aucta; e. julus perfectus, f. ejus calyx persistens; g. calyx separatus a facie interiore; h. julus separatus et per longitudinis medium dissectus, i. ejus axis; k.k. stamina aliquot separata; l. stamen in magnitudine aucta cum filamento m perbrevi, anthera n. magna et o. ejus extremitatis squama; p. anthera transverse dissecta; q. julus aridus; r. stamen a facie superiore, s. in magnitudine aucta, t. a facie laterali, u. inferiore, x. in magnitudine aucta, y. pars pollinis quem continet; 1. conus junior, 2. ejus calyx; 3. calyx separatus a facie externa, 4. interna; 5. ejusmodi conus per longitudinis-medium dissectus, 6 ejus axis, 7. ejus pars inferior, 8. axis denudatus; 9. squamula separata a facie inferiore, 10. superiore; 11. conus paullo adultior; 12. coni maturi pars inferior, cujus squamis semina incumbunt; 13.*

Humbaba the guardian he smote to the ground,
for two leagues afar…
With him he slew…
the woods he.…

He slew the ogre, the forest's guardian,

at whose yell *were sundered* the peaks of Sirion and Lebanon,…
the mountains did quake,…
all the hillsides did tremble.

He slew the ogre, the cedar's guardian,
the broken.…
As soon as he had slain all seven (of the auras),
the war-net of two talents' weight, and the dirk of eight,

a load of ten talents he took up, he went down to trample the forest.
He discovered the secret abode of the gods,
Gilgamesh felling the trees, Enkidu choosing the *timber*.

Enkidu opened his mouth to speak, saying to Gilgamesh:
"My friend, we have felled a lofty cedar,
whose top thrust up to the sky.

"I will make a door, six rods in height, two rods in breadth, one
cubit in thickness,
whose pole and pivots, top and bottom, will be all of a piece."

… he went trampling through the Forest of Cedar,
he discovered the secret a bode of the gods.
The Wild-Born knew how to give counsel,
he said to his friend:

"By your strength alone you slew the guardian,
what can bring you dishonour?
Lay low the Forest of [Cedar!]
Seek out for me a lofty cedar,
whose crown is high as the heavens!" (*Gilgamesh*)[50]

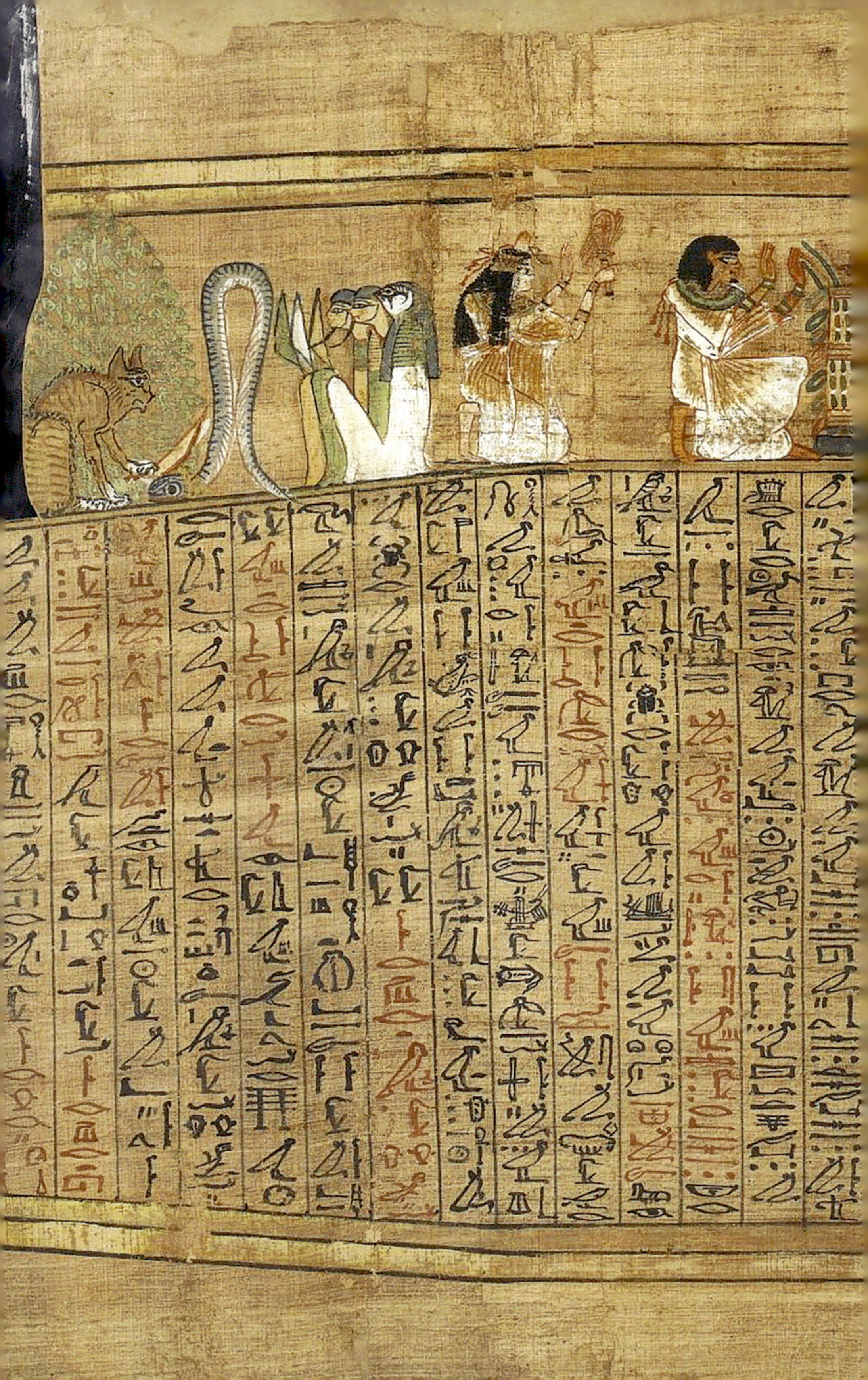

"Two great Persea trees have grown, as a great marvel of his majesty, in the night by the side of the great gate of his majesty." And there was rejoicing for them in all the land, and there were offerings made to them.

And when the days were multiplied after these things, his majesty was adorned with the blue crown, with garlands of flowers on his neck, and he was upon the chariot of pale gold, and he went out from the palace to behold the Persea trees: the princess also was going out with horses behind his majesty. And his majesty sat beneath one of the Persea trees, and it spake thus with his wife: "Oh thou deceitful one, I am Bata, I am alive, though I have been evilly entreated. I knew who caused the acacia to be cut down by Pharaoh at my dwelling. I then became an ox, and thou causedst that I should be killed."

And many days after these things the princess stood at the table of Pharaoh, and the king was pleased with her. And she said to his majesty, "Swear to me by God, saying, 'That which the princess shall say to me I will obey it for her.'" And he hearkened unto all she said. And he commanded, "Let these two Persea trees be cut down, and let them be made into goodly planks." And he hearkened unto all she said. And after this his majesty sent skilful craftsmen, and they cut down the Persea trees of Pharaoh; and the princess, the royal wife, was standing looking on, and they did all that was in her heart unto the trees. But a chip flew up, and it entered into the mouth of the princess; she swallowed it, and after many days she bore a son. And one went to tell his majesty, "There is born to thee a son." And they brought him, and gave to him a nurse and servants; and there were rejoicings in the whole land. And the king sat making a merry day, as they were about the naming of him, and his majesty loved him exceedingly at that moment, and the king raised him to be the royal son of Kush.

Now after the days had multiplied after these things, his majesty made him heir of all the land. And many days after that, when he had fulfilled many years as heir, his majesty flew up to heaven. And the heir said, "Let my great nobles of his majesty be brought before me, that I may make them to know all that has happened to me." And they brought also before him his wife, and he judged with her before him, and they agreed with him. They brought to him his elder brother; he made him hereditary prince in all his land. He was thirty years king of Egypt, and he died, and his elder brother stood in his place on the day of burial. (*The Tale of Two Brothers*)[51]

猿わか町秋葉の境内

This year Kahabe no Omi was sent to the province of Aki with orders to build ships. On arriving at the mountain, he sought for ship timber. Having found good timber, he marked it and was about to cut it, when a man appeared, and said:—

"This is a thunder-tree, and must not be cut." Kahabe no Omi said:— "Shall even the Thunder-god oppose the Imperial commands?" So having offered many mitegura, he sent workmen to cut down the timber. Straightway a great rain fell, and it thundered and lightened. Hereupon Kahabe no Omi drew his sword, and said:—"O Thunder-god, harm not the workmen; it is my person that thou shouldst injure." So he looked up and waited. But although the God thundered more than ten times, he could not harm Kahabe no Omi. Then he changed himself into a small fish, which stuck between the branches of the tree. Kahabe no Omi forthwith took the fish, and burnt it. So at last the ships were built. (*The Nihongi*)[52]

Imagination is the beginning of creation.
You imagine what you desire, you will what
you imagine and at last you create what you will.
—George Bernard Shaw,
Back to Methuselah

EPILOGUE

Imagination and Beyond

THROUGH IMAGINATION we form our understanding of what lies beyond our immediate senses. Imagination is an act of mental creation that helps to form the world in which we live. The prevailing human imagination of plants is limited to viewing them as resources to meet our needs. This is as much a function of our philosophical tradition and mythology as it is of the fact that the sensory apparatus of plants is not readily visible to the human eye.

I hope that the botanical myths contained here challenge readers to re-imagine plants as fully alive, as having their own purposes, as deserving of respect, as our sentient relations. This re-imagination of the plant kingdom is integral to Val Plumwood's call to re-situate nature within the moral sphere for the way in which we *imagine* plant life shapes the way in which we humans *behave* toward the natural world.

Imagination, though, is just the beginning. The myths contained here give us a glimpse of a world beyond our immediate senses. This experience challenges the view of plants as passive and inert. It allows us to stretch our minds beyond and through the ontological differences between humans and plants, to see and feel the continuities between plant and animal bodies and to recognize the kinship connections that undercut the is/ought gap (or fact/value divide) through which has grown a skepticism toward treating plants ethically. However, in order to really bring plants (and nature) within the moral sphere we need to move from our imagination and into our will, from our ideas and into action.

The first step must be to take inspiration from mythological stories and to use and retell these tales. Readers from both within and without the traditions represented must do what artists, poets, and writers have done for thousands of years. Re-telling, re-shaping, and expanding lineages of stories in which plants are depicted as sentient and fully alive is critical. For those within cultures with strong botanical mythology, this may well be a case of continuing to keep these myths alive. For those without, such as myself, this is a task of taking inspiration from the myths in this collection, and without appropriating or misrepresenting them, transforming this inspiration into new stories of the world of plants.

This is something that has been happening for centuries in children's literature. Childhood tales, such as Aesop's *The Oak and The Reed*, Hans Christian Andersen's *The Fir Tree* and *The Last Dream of the Old Oak Tree*, or Shel Silverstein's *The Giving Tree*, are part of the culture of stories needed to recast nature as sentient and mindful. If this work can provide fresh inspiration for stories of such quality (particularly in cultures that currently lack a living botanical mythology) it will have been worth the paper it was written on.

However, the "stories about something significant" are just one aspect of mythological tradition. Myth itself is *part* of a culture[1] and is entwined with symbolism and ritual. Indeed, myth and ritual are commonly seen as two aspects of the "religious" practice in traditional societies. The exact way in which myth and ritual connect has been debated and disputed by scholars. Where Tylor saw rituals as arising from myth, Burkert saw them as independent, and Eliade saw myths as an explanation for rituals.[2] I am not concerned with pinning down the precise relationship between myth and ritual (or a definition of ritual),[3] but more with exploring how ritual may help us move from the narrative of myth into the corporeal world of bone, bark, and blood. Tom Driver's account of ritual provides an excellent platform for these purposes:

> Ritual belongs to what the ancient Greeks would have called technê, the root of our words technical, technique, and technology. Ritual is a sort of technology because it is a method (a time-honored one) for accomplishing something in the real world.... The technê of ritual, however, is utterly different from modern technology. Its field of action is not an objectified physical world but a divine, human, animal, and vegetative cosmos of mores, moralities, and mutual relationships.[4]

Regardless of the exact relationship between myths and ritual, mythological stories exist in a cultural context that locates and embeds them in the real world. Now that we have drawn inspiration from botanical myths we must look to take this wider contextual approach and embed the stories of plants as sentient kin in our morality and in our relationships. We need to go one better than Charles Darwin, who recognized our theoretical kinship with other species, and recreate our actual kinship relationships with plant life. Our field of action then is human activities, behaviors and institutions.

Suggestions abound for how to approach this task. At an institutional level, the legal scholar Christopher Stone has put forward the idea that our trees need standing in the legal system, replacing the restricted imagination of them as purely property, with a respect for the life of trees via legal standing.[5]

The scholar of ritual Ronald Grimes suggests that we look to ritualistic performances, such as kissing the earth beneath our feet, or circumambulating trees to move our idea of respect for nature into our behaviors in order that we may earn "the respectful silence of carrots or the applause of vultures."[6] Pagan writers have suggested that human beings ask before they take from food and other stuffs from plant life—perhaps taking influence from Indigenous peoples, such as Māori who seek permission and offer placation.[7] Such a simple act helps to disrupt the notion that plants are there purely for our sake.[8] This approach also echoes that of Indigenous peoples who advocate leaving behind plant foods so that the plants and other animals may also live, and connects with the wisdom of Indigenous leaders such as Bill Neidjie who not only recommended speaking with the plants but listening to them too.

> That tree now, feeling …
> e blow …
> sit quiet you speaking …
> that tree now e speak …
> that wind e blow …
> e can listen.[9]

Taking time to listen to our sentient plant kin will be key in our attempts to change our attitudes and behavior toward them. By listening, I don't necessarily mean trying to discern audible sounds from the bodies of plants. I am attracted by Fernando Pessoa's description of the "audible smile of the leaves" as much as I am fascinated by the studies that show that plants are capable of making, and responding to, and differentiating between sounds. But, Bill Neidjie's reference to listening is much more fundamental, more integral than a reference to something audible.[11] There is something about the way he phrases the line "that tree now e speak" which suggests to me that the trees should be allowed to be at the forefront of activity, almost that they should be allowed to take *their* turn at describing, shaping, and living in the world. Human listening then involves stepping back, ceasing our endless commentary, putting aside our own perspectives, and allowing plants the philosophical and physical space they need to flourish.

BOOK OF GENESIS

The repetitive, formulaic account of cosmological creation in Genesis 1 is the later of the two biblical accounts, with its origins around 500 BCE. Scholars have long highlighted its strong similarities with (and likely origins in) the Babylonian creation tale *Enuma Elish*.[1] The sequences of both stories are indeed remarkably similar. From a primeval darkness, first light is created, followed by the firmament, dry land, the luminaries, and then man—the creator God(s) then rest(s).[2] In the *Enuma Elish*, humankind is created for the service of a plurality of gods, whereas in Genesis 1 humans are created in the image of the one God, and the animals and plants of the Earth are created in service of humankind. The combined narrative argues against the Mesopotamian theory of creation and is "compelling in its archetypal character, its adaptation of myth to monotheistic ends."[3]

The second account of creation, in Genesis 2, is around five hundred years older, composed in approximately 1000 BCE. In this myth, the action is more personal and earthy, a fact reflected in the use of the personal name of God, Yahweh, rather than the more formal *elohim* (as used in Genesis 1). This grounded account is also reflected in a narrative that deals more with the travails of human life on Earth. Genesis 2 also has close similarity with the *Atra-Hasis* epic, another ancient Mesopotamian tale from eighteenth century BCE, including the situation of a divine garden and the exploration of the relationship between humankind and the natural world.[4]

THE EDDAS

The *Gylfaginning*, or the *Beguiling of Gylfi*, is the first of three texts in Storri Sturlison's famous medieval compilation of Old Norse mythology, *The Prose Edda* (c.1220). The *Gylfaginning* tells the story of Gylfi, a king who "ruled the land that men now call Sweden."[5] Gylfi is tricked by one of the goddesses of Æsir and travels to Asgard to find out how all Æsir people bend "all things according to their will."[6] On the way he is tricked again, and ends up in a great hall, where he introduces himself as Gangleri. There he is challenged to ask questions of three men, including a king named Hárr (one of the two hundred–plus names of the chief Norse god, Odin), in order to prove his wisdom. Gangleri (Gylfi) asks questions about the gods and about the creation and destruction of the world, to which the three men provide detailed answers.

Much of the material in the *Gylfaginning* (and the entire *Prose Edda*) draws on older, poetic material, found in a medieval manuscript known as the Codex Regius of the elder Edda, particular the Old Norse poem *Völuspá*.[7] This collected material has come to be known as the *Poetic Edda*. Both Eddas are major sources of Old Norse mythology, yet it is important to note that both were written by Christians more than two hundred years after the official recognition of Christianity in Iceland (c.1000 CE).[8] The material in the Eddas is therefore a presentation, or an interpretation of ancient pagan mythologies. What is certain is that the myths within would originally have been crafted to be told, and performed, orally rather than being read in private.

The provenance of the myths in the Eddas are the subject of scholarly debate. Much of the material has its origins outside Iceland, where the Codex Regius manuscript was compiled.[9] Many of the plants mentioned in the myths are testament to these older origins. For example, fir, oak, and ash trees (think of the world tree Yggdrasil) did not, and do not, grow in Iceland.[10] It is likely, then, that the oral traditions set down by Icelandic scholars themselves originate from outside Iceland.[11]

THE VEDAS

The Vedas (the word *veda* means knowledge) constitute the four foundational texts of Hinduism, the *Rig Veda*, the *Sama Veda*, the *Yajur Veda*, and the *Atharva Veda*.

The *Rig Veda* is a vast collection of 1,028 hymns and 10,600 verses, composed in an archaic Vedic Sanskrit between 1500–1200 BCE, making it the oldest of the Vedas.[12] Like many of our mythological sources, it is not strictly a text, having originally been composed and transmitted orally. Yet, unlike other sources, this oral transmission was very rigid, often done without the change of a single word.[13] The *Rig Veda* is therefore one of the oldest and best-preserved sources of mythology in the world.

The archaic nature of the Sanskrit language, and the lack of contemporaneous texts, has rendered the *Rig Veda* difficult to translate and interpret. Most of the Vedic texts were "composed by Brahmins for Brahmins"[14] and the hymns in the *Rig Veda* often emphasize obscure ritual practices, making them difficult for modern minds to fully comprehend. While they are venerated, the subject matter is rarely engaged with directly by modern Hindus. Scholar Frits Staal rightly cautions against reading modern theories into the *Rig Veda*,[15] yet there are well-known linguistic and cultural connections between the *Rig Veda* and the Iranian *Avesta*, including references to the sacred plant *soma/haoma* from the proto-Indo-Iranian *sauma*[16] and creation following the death of the first being.[17] The *Rig Veda* contains a significant collection of hymns dedicated to various deities. Mandala nine comprises hymns that are dedicated to *soma*.

The *Atharva Veda* (named after Atharavan, a Vedic priest) is comprised of a series of myths, incantations, prayers, and hymns.[18] Scholarship indicates that it is probably the youngest of the four vedas.[19]

BUNDAHISN

The *Bundahisn* (meaning "original creation") is a collection of Zoroastrian cosmogony, mythology, and legendary history composed in the eighth/ninth centuries CE, in the exclusively written Middle Iranian language known as Book Pahlavi.[20] The *Bundahisn* is fragmentary, with numerous omissions and dislocations. The two surviving versions (recensions) differ in both their arrangement and their length.[21] The work draws upon the *Avesta* (the primary Zoroastrian texts) and many passages appear to be translations from an *Avesta* original.[22]

The *Bundahisn* is unlikely to have been the original name for this work. Translator E. W. West regards the name to have originated from the repetition of the word within the first sentence of one recension, and from its appropriateness to the subject matter, which, at least in its early parts, deals with the progressive development of creation under a range of good and evil influences.[23] West asserts that the same sentence tells the name of the text as *Zand-Ākāsīh*, or "knowing the tradition."[24]

The cosmogony and mythology in the *Bundahisn* is rich and detailed. In the beginning, a good spirit and an evil spirit do battle over their respective creations. The evil spirits' conflicts with the six classes of creation (sky, water, earth, plants, animals represented by the primeval ox Evagdad, and humans represented by the primeval human Gayomard) cause these creations to wither and eventually die. Following this death, the text describes many aspects of the created world, such as the regions, the mountains and sea, the classes of animals, the rivers and lakes, and different kinds plants, including the Gaokerena (also known as the Gokard or White Hom tree).

The myth of Pangu first appears in the second century at the time of the Three Kingdoms, in Xu Zheng's *Sanwu Liji* and *Wuyun Linian Ji* (222–280 CE) and is one of very few Chinese myths that tell of the origins of the cosmos.[25] The story begins in a time of chaos and tells of the separation of heaven and earth through Pangu's birth, and the creation of the natural world through his death. Many scholars have asserted that, because of its strong similarities with Indo-European creation myths in which the death of a first being creates the world (e.g., the primeval ox, Evagdad), and the fact that the myth is unknown in China before the second century, it is Western Asian/Indian in origin.[26]

However, in addition to written sources, the figure of Pangu also exists in the mythology of southern minority groups in China. Wu points to this fact, along with archaeological and linguistic evidence of the existence of the Pangu myth before the second century, and argues for the origin of the myth within China.[27] The myth of Pangu also still plays a role in contemporary Chinese society. Southern minority groups such as the Miao, the Yao, and the She in Hunan and Guangxi still place a tablet for Pangu in the place where they worship their ancestors.[28]

THE *KALEVALA*

The *Kalevala* is a work of epic poetry, composed by Elias Lönnrot from his collections of traditional songs and poems from the Karelia region of eastern Finland. First published in 1833–35 as the *Old Kalevala*, Lönnrot continually reworked the material into further versions, including the most commonly read version, published as the *Kalevala* in 1849.

Väinämöinen is the central figure of the *Kalevala*. The text begins with his birth upon the water, and his floating upon the seas for nine years until he reaches land. Many of the runos (or cantos) in the *Kalevala* tell of Väinämöinen's adventures, including his perpetual search for a wife, and his actions which help create the world, such as the summoning of a man to fell the great oak.

Folklorists have long debated the authenticity of the material in the *Kalevala*. The key concern is whether the extensive editing, rearranging, and rewriting of the original material has rendered the *Kalevala* an inauthentic literary epic of Lönnrot's creation, rather than an authentic collection of Karelian poetry. Critics have pointed out the strong influence of romanticism, and other epic poetry, including the *Völuspá* from the *Poetic Edda*, and argue that the text is disconnected from the historical and cultural reality of the folk poetry it draws on.[29] Others have defended Lönnrot's creative vision as an authentic retelling of Karelian mythology.[30]

The *Kalevala* is the national epic of Finland and has played a significant role in the development of Finnish national identity, which culminated in Finland's independence from Russia in 1917. The themes and characters from the *Kalevala* have had an extensive impact on Finnish society, influencing the names of places and people, politics, fine art, literature, and popular music.[31] The national epic even has its own holiday, Kalevala Day, celebrated on February 28 each year, commemorating the date on Lönnrot's foreword in the first version of the *Kalevala*.[32]

MAHĀBHĀRATA

The *Mahābhārata* is an ancient Indian epic, compiled by the fourth century BCE and reaching its final form in the fourth century CE. The main narrative concerns the Kurukshetra War and the battle between the Kaurava princes and their cousins the Pandavas. At 200,000 verse lines and 1.8 million words, the *Mahābhārata* is the longest epic poem ever written, roughly ten times the length of Homer's *Odyssey* and *Iliad* combined. The text is enormously complex, with a structure which has been labeled by scholars as a "monstrous chaos."[33]

In addition to the main narrative, the *Mahābhārata* also contains substantial historical, devotional, and philosophical material, and is considered by Hindus to be sacred scripture. The *Mahābhārata* also contains texts considered to be works in their own right, such as the *Bhagavad Gita*. The philosophical discussion about the nature of duty and action (karma) between Arjuna and Lord Krishna in the *Bhagavad Gita* elicits many of the principles of Hinduism and remains profoundly influential on Indian society today. Along with the other great Indian Sanskrit epic, the *Ramayana*, a tradition of scholars has also examined the mythical basis of the *Mahābhārata*.[34] Links have been explored between key characters in the text and various mythic phenomena including Hindu gods and celestial objects, as well as links with other Indo-European mythical sources such as the Eddas.[35]

THE PURANAS

The Puranas are an encyclopedic genre of ancient Indian literature (in Sanskrit the name *Purana* literally means "ancient" or "old") found in both Hinduism and Jainism.[36] The name *Purana* implies the preservation of ancient traditions, but

Wilson, the first translator of the *Vishnu Purana*, noted that this was a purpose "very imperfectly fulfilled"[37]—it being well known that the texts themselves are inconsistent, including marked differences among multiple manuscript versions of the same Purana.

In Hinduism, the Puranas are not considered to be scripture. Unlike other texts, the Puranas were accessible by women and lower castes, thereby giving "a portrait of popular religiosity through the ages."[38] However, many are named after Hindu deities, including the *Vishnu Purana*, one of the most celebrated of the eighteen Mahapuranas (Great Puranas) and the first to be translated from Sanskrit into English.[39] The Puranas cover a vast array of topics, as diverse as medicine, governance, philosophy, and pilgrimage routes, but many have significant cosmogonical and mythological content.

The *Vishnu Purana* is named after Vishnu, one of the principal deities of Hinduism. Wilson noted that the text repeats "the theoretical cosmogony of the two great poems,"[40] the *Mahābhārata* and the *Ramayana*. The origin of the earth is covered in the second part of the text, including the creation of the continents, oceans, rivers, and the plants that clothe the earth's surface. The text is part of the literature of Vaishnavism, a major tradition within Hinduism that considers Vishnu as supreme among the Hindu deities. In Vaishnavism, Vishnu is equated with the concept of Brahman, the underlying, ultimate reality in the universe. Vishnu has many avatars, or incarnations, including Krishna in the *Mahābhārata*. The longest section of the *Vishnu Purana*, dedicated to the story of Krishna, is very similar to that found in the *Bhagavata Purana*.[41]

The *Padma Purana* is named after the sacred lotus flower in which Brahma was born (note the link to the myth of Padmasambhava's birth) and is one of the largest of the eighteen Mahapuranas at 55,000 verses long. The text is a large and diverse compilation of topics including cosmology, the origins and deeds of deities, philosophical discussions on the soul, the geography of rivers and regions, and Indian pilgrimage sites. The surviving manuscripts of the *Padma Purana* exist in two major versions, and within these groups, different language versions show major inconsistences.[42] The Hindu *Padma Purana* is not to be confused with a Jaina text of the same name.

POPOL VUH

The *Popol Vuh* ("Book of the Community," "Book of the Council," or more literally "Book of the People") is a Quiche Mayan text of mythology, cosmogony, and genealogical history.[43] Using genealogical references in the text, some scholars have suggested that the *Popol Vuh* was written between 1554 and 1558, shortly after the Spanish conquest of Guatemala.[44] Others have questioned the existence of a written Quiche text, suggesting an oral recitation as the source material.[45] In the early 1700s, a Dominican friar, Father Francisco Ximénez, carefully transcribed the original Quiche material from his parish at Santo Tomás Chichicastenango and translated it into Spanish.[46] Around 1714, Father Ximénez incorporated this material into a Quiche/Spanish manuscript with the title *Historias del origen de los Indios de esta Provincia de Guatemala*.[47] This manuscript remained in obscurity until 1947 when Adrian Recinos published a Spanish translation of the Quiche from Father Ximénez's text.[48]

The *Popol Vuh* details the "cosmogonical concepts and ancient traditions" of the Quiche Mayans, "the history of their origin, and the chronology of their kings down to the year 1550."[49] Scholars have long been aware of the obvious influence of the Bible on the creation stories in the *Popol Vuh*.[50] The opening account of the creation of the Earth contains passages that are almost direct transcriptions from the Book of Genesis and the text itself acknowledges the fact that the *Popol Vuh* was written in light of the "Law of God" and Christianity.[51] Despite this clear colonial presence, and the fact the source of Father Ximenez's writings remain unknown, the *Popol Vuh* is regarded as a hugely important text, a rare example of Mesoamerican mythology.

The *Books of Chilam Balam* (BCB) are handwritten manuscripts written in the Mayan language (using adapted European script), which contain a diverse array of material including fragments of historical narratives, rituals, medical treatises, and mythological accounts of the creation of the world.[52] The name *Chilam Balam* is derived from a pre-Hispanic oracle (*chilan*) from the town of Maní in the northwestern Yucatán with the surname Balam or "Jaguar," who it is said prophesied the arrival of the conquistadors and Christianity.[53] The *BCB of Chumayel* is one of nine extant Maya BCB (Chan Cah, Chumayel, Ixil, Kaua, Na, Tekax, Tizimin, Tusik, and Códice Pérez) that are named after the towns in the Yucatán, Mexico, in which they were found.[54]

The nine extant BCB primarily date to the eighteenth century, but many passages are thought to have been transcribed from older hieroglyphic manuscripts, still in existence in northern Yucatan in the late 1600s.[55] Seized by religious zealotry, the Spanish burned many of the codices written in Mayan hieroglyphic texts, rendering the surviving BCB a vitally important source of Mayan history, culture, literature, and mythology. Knowlton compares the relationship of the BCB and Mayan civilization to the Eddas and our knowledge of pre-Christian Norse culture.[56]

The myths in the *BCB of Chumayel* include the creation of the world and the origins of plants, animals, and humankind. Like the *Popol Vuh*, the *BCB of Chumayel* is a postcolonial text, composed within a strong Christian context, raising scholarly debates about the authenticity of the Mayan contents. There are obvious Biblical influences, but most of the content is Mayan,[57] with its roots likely in ancient Maya civilization.[58] Regardless, to my mind, such claims of "inauthenticity" misunderstand the nature of story and myth which is never static and fixed, but is continually being "reworked nostalgically and adapted creatively to the present."[59]

ACOMA

Acoma Pueblo is the westernmost of the Keresan language–speaking *pueblos* (*pueblo* is the Spanish word for town or village) in New Mexico. More than eight hundred years old, it is one of the oldest continuously occupied communities in the United States.[60]

The Acoma origin myth excerpted here was collected by Matthew Stirling, the chief of the Bureau of American Ethnology, when a group of Acoma people visited the bureau in August 1928, and was first published in December 1930 by Daryll Forde. A similar version of the story was also collected by a colleague, Leslie White, around the same time. Forde notes the similarities with other Keresan language creation myths, particularly Sia and Laguna, but "marked differences in incident and names occur."[61] The creation myth of Tsichtinako takes place almost entirely on Earth, and not in the Underworld, but Tsichtinako's baskets parallel the parcels of the Spider Woman (Sus'sistinako) of Sia mythology.[62]

YANYUWA

The Yanyuwa are an Indigenous Australian people whose country includes the Sir Edward Pellew Group of Islands, the Bing Bong area, and the delta regions of the McArthur River and Wearyan River in the southwestern Gulf of Carpentaria, Northern Territory.[63] Yanyuwa have been dispossessed from their lands for many years. Some country has been returned, but most Yanyuwa live in and around the township of Borroloola.[64]

The excerpted tale of the Tiger Shark ancestor (Yulungurri) is a Dreaming story about the origins of one of the traditional staple foods in Yanyuwa country, kernels of the cycad (*Cycas angulata*), told to the anthropologist John Bradley as part of his three decades of collaborative work with Yanyuwa people.[65] Naturally highly poisonous, cycad kernels are at first glance a strange food source. However, with some meticulous, labor-intensive processing, Yanyuwa made them not only edible, but a cherished food that is now sadly no longer part of an everyday diet.[66]

The story of the cycad and the processing of its fruits helps construct the relationships between Yanyuwa, their plants, and their country. Yanyuwa people describe themselves as li-Anthawirryarra, generally translated as "people of the sea."[67] Brought to country by the Tiger Shark, the cycad also has its origins in the sea—the history of Yanyuwa people and the cycad are intertwined.

WARRAMURRUNGUNDJI

Warramurrungundji is a female ancestral being who distributed food plants around country in Arnhem Land, Northern Territory, Australia. The tribes and clans of Arnhem Land trace the travels of Warramurrungundji from the sea near the northeast of the Cobourg peninsula and across their various countries.[68] Warramurrungundji created human beings, designated languages, brought plants, formed rivers and waterways, and at the end of her journey transformed herself into a rock.[69]

The myth of Warramurrungundji, excerpted here, comes from an unpublished 1980s ethnobotanical study by George Chaloupka of Nadjolaminj estate, now part of Kakadu National Park.[70] The study includes information on creation myths and the material uses of plants, for foods, cultural items, and poison of the Badmardi clan, Mayali language speakers.

Mayali is a term for languages spoken in western Arnhem Land, Northern Territory Australia, but there is some confusion around its use. Some academics use Mayali as a cover term for a dialect chain encompassing Gundjeihmi, Kunwinjku (Gunwinngu), Kundedjnjenghmi, Kuninjku, and Manyallaluk Mayali.[71] Yet, some Kunwinjku speakers do not accept this overarching term, confining "Mayali" as a descriptor of Gundjeihmi, Kundedjnjenghmi, and Manyallaluk Mayali. The term *Bininh gunwok* is used as an alternative overarching term, from *bininj* "Aboriginal person" and *gun-wok* "language."[72]

John White's *The Ancient History of the Maori* (1887–1890) is a six-volume work on virtually every aspect of Māori knowledge, history, and tradition, including Pūrākau (myths), whakapapa (genealogies), whakatauki (proverbs), karakia (prayers), and waiata (songs).[73] All geographic and tribal areas of Aotearoa are represented, though the most significant tribal source (by quantity) is Ngāi Tahu.[74] Commissioned by the government, the work was originally intended to be twenty volumes, but was canceled after the first six following political interference, financial cutbacks, and poor administrative oversight.[75]

Along with George Grey's *Polynesian Mythology*, *The Ancient History of the Maori* is part of a tradition of nineteenth-century Pākehā scholars, "who edited, translated, and interpreted Māori traditions, not always for the best or in ways appropriate to a Māori world view, but establishing a local tradition of Pākehā interest and respect for Māori knowledge."[76] White's methods of collecting and editing Māori traditions have been controversial, with accusations of poor scholarship and literary fabrication leveled from the outset.[77] The sheer scale of White's work has meant that *The Ancient History* "remains an important repository of Māori language and learning," but a significant amount of scholarship has been generated "attempting to disentangle the authentically Māori sources from his own interpolations to them."[78] The story of Rātā, excerpted here, shares strong commonalities with other versions recorded elsewhere, indicating that whatever augmentations White inserted, these myths were known to pre-European Māori in a similar form.[79] However, the story of Tāne bringing the trees from heaven is only known from White, and so the status of this myth is not as certain.

NIHONGI

The *Nihongi* (translates as *Japanese Chronicles*), otherwise known as the *Nihon Shoki* (*The Chronicles of Japan*), is the second-oldest book of Japanese history after the *Kojiki*. The *Nihongi* is an invaluable source for Japanese history, as well as mythology, poetry, political structures, and social customs. The early commentary on the *Nihongi* (the *Konin Shiki*), regards the text as a compilation of much older material, detached passages linked together in a narrative sequence.[80] The Introduction to Aston's translation relates that it was completed in 720 CE by Prince Toneri and Yasamaro Futo no Ason (Futo no Yasumaro).[81]

The text was written in classical Chinese, a common occurrence for official documents at the time. William George Aston, the early translator into English, notes the clear intrusion of Chinese philosophical and cultural ideas into the earlier sections of the work.[82] Regardless, the early sections of the *Nihongi* contain a version of the Japanese creation myth, which along with the version in the *Kojiki*, forms part of the literary basis for Shinto.

LIFE OF THE BUDDHA

Siddharta Gautama, the Indian ascetic known as the Buddha (or Buddha Sākyamuni), lived in present-day Nepal during the late sixth and early fifth centuries BCE.[83] There are countless accounts of the life of the Buddha, the first of which was set down around 80 CE in the Pali Canon in Sri Lanka.[84] André Ferdinand Herold's 1922 biography, *The Life of the Buddha* (translated by Paul C. Blum), lays down a simple and nontechnical narrative of Siddharta Gautama's life, drawing from the several classic Buddhist texts, particularly the *Buddhacharita of Asvaghosa*, the *Lalita-Vistara*, and the *Jataka*.[85] The biography charts the well-known story of a young man, born into luxury, who seeks and finds enlightenment at Bodh Gaya beneath the Bodhi tree, a large fig tree species, *Ficus religiosa*.

PLINY, *NATURAL HISTORY*

Pliny the Elder's *Natural History* (*Historia Naturalis*) is an encyclopedic work covering an array of topics, including traditional "natural history" subjects such as botany and zoology, but also wider topics, including agriculture, horticulture, geology, anthropology, astronomy, and mathematics. The work is composed of ten volumes divided in thirty-seven books, with most of the botanical material contained in books 12–27. Having begun the work in 77 CE, Pliny was still revising the *Natural History* at the time of his death in Pompei following the eruption of Vesuvius in 79 CE.[86]

The *Natural History* is perhaps the most influential natural history work ever written.[87] For the 1,400 years from Pliny's death to the publication of the first print edition in 1469, the *Natural History* was one of the principal sources of botanical knowledge, along with Dioscorides's *De Materia Medica*.[88] For nearly three hundred years following the first print edition, the *Natural History* was widely reprinted across Europe and cited as the authority by students of natural history.[89]

Pliny's approach to writing the *Natural History* was to compile numerous facts about the topics at hand from nearly five hundred different authors, being careful to cite the source of information. Pliny's principal source is undoubtedly Aristotle, and his view of plant life echoes Aristotle's hierarchical ordering of the natural world. However, Pliny also included sources and information that scholars have regarded as hearsay.[90]

DODONA

Dodona is the site of the oracular shrine of Zeus, in northwestern Greece. There, the rustling (and/or the speech) of the ancient oak trees helped to determine the correct course of action to be taken. Dodona is reportedly the oldest of the oracular shrines—one that has had a prolonged effect on ancient Greek culture. From a first mention in the *Iliad*, there are repeated references to Zeus and Dodona throughout ancient Greek literature.[91] Plato's dialogue *The Phaedrus* is then one of a number of ancient Greek sources for the story of the Dodonan oaks, along with Hesiod's *Catalogues of Women Fragment* and Herodotus's *Histories*.

KOJIKI

Completed in 712 CE, the *Kojiki* (translated as *Records of Ancient Matters*) is a supremely important source for its records of the ancient history, language, and culture of Japan. Although written in Chinese characters to represent the Japanese language (a form of writing known as Man'yōgana), the Chinese influence is less than on other ancient Japanese texts and the presentation of ancient culture is regarded to be authentic.[92]

The *Kojiki* is prized for its wonderful storytelling, despite being heavily footnoted in the manner of an academic text, and is regarded as being a more approachable text than the *Nihongi*.[93] The *Kojiki* was compiled by Futo no Yasumaro, who is also credited with involvement in compiling the *Nihongi*. Along with the *Nihongi*, the work is one of the two primary texts for Shinto, the national religion of Japan.

Like the *Nihongi*, the opening parts of the *Kojiki* deal primarily with cosmology and mythology. The *Kojiki* details the creation of the Japanese islands and the births of innumerable deities are intermingled with the tales of their deeds, including the bringing of the sacred saka-ki tree from Mount Kagu, one of the three mountains of Yamato, in Kashihara, Nara Prefecture. The text then moves onto a number of historical legends and ends with a detailed chronology and stories of the early Imperial line.

ETHNOBOTANY OF THE ZUNI

Zuni are a tribe of Native American people who predominantly live in the Zuni River Valley, a tributary of the Little Colorado River, New Mexico.

The story of A'neglakya and A'neglakyatsi-tsa is taken from a paper on Zuni ethnobotany by pioneering ethnologist Matilda Coxe Stevenson, published in the 1908–09 Annual Report of the Bureau of American Ethnology.[94] Stevenson was on the staff of the bureau and made detailed studies of all the Pueblo tribes of New Mexico, including the Zuni. She first visited the Zuni Pueblo in 1879 and soon began collecting information on plant uses, taxonomy, and mythology, accompanied by voucher specimens.

In addition, Stevenson also collected information about the relationships between the Zuni and their plants, noting that the Zuni speak with their plants, acknowledge the substantial kinship relationships between humans and plants, and perceive plants as sentient beings.[95] Plants are regarded as the gifts of Earth Mother, one of the three main deities in Zuni religion.[96] Stevenson notes that symbols of these relationships are incorporated into fabric designs, ceramics, and ceremonial objects.[97]

FABLES AND RITES OF THE INCAS

Cristoval de Molina was a Spanish priest at the Hospital for the Natives of Our Lady of Succour, in Cuzco, Peru, from 1565 to his death in 1585. His manuscript *Account of the Fables and Rites of the Incas* (*Relacion de las fabulas y ritos de los Incas*) is a report to Dr. Don Sebastian de Artaun, the bishop of the ancient Inca capital. The report is not dated, but as Artaun was bishop of Cuzco between 1570 and 1584, it must have been written during this period.[98] The text was translated into English by Clements Markham in 1873. The original manuscript is in the Spanish National Library in Madrid.[99]

An excellent Quechua speaker, Molina was well acquainted with the daily lives of the Inca through his work at the hospital, and this allowed him to capture their rituals and religious celebrations in almost unprecedented detail.[100] Molina's manuscript provides a detailed account of the ceremonies performed by the Incas at different times of the year, with descriptions of the prayers, rituals, and festivities. It also includes a chapter with a lengthy description of the Inca origin myths.

Molina's manuscript is based on the testimonies of the older Inca men of Cuzco, the last surviving generation who witnessed the rituals and religious practices before Spanish conquest.[101] Written at a time of rapid social and cultural change, with increasingly aggressive Spanish attempts to wipe out what were seen as idolatrous practices, Molina's manuscript is a tremendously important record of Inca culture.[102]

VIRGIL, *AENEID*

Virgil's masterpiece, the *Aeneid,* is a Latin epic poem that tells the story of Aeneas, his flight from Troy to Sicily, and his leadership of the Trojans to war in Latium (now the region of Lazio around Rome). The poem ends with Aeneas founding the city of Rome.

Composed by the Roman poet Virgil between 29 and 19 BCE, the *Aeneid* was entirely conceived and composed in writing. Before it was complete Virgil intended to journey in Greece and the eastern Mediterranean to revise the manuscript.[103] At Athens, he was convinced by the Emperor Augustus to return with him on his homeward journey, contracted fever, and died just after returning home.[104] Before his death, Virgil requested friends to burn the manuscript of the *Aeneid* due to many perceived imperfections, but Augustus ordered its publication.

One of the key classical texts, studied and memorized for nearly two thousand years, the *Aeneid* has strongly influenced the development of Western art and literature. Dante's *Divine Comedy*, itself a cornerstone of Western literature, shows some of the strongest influence, with the text proudly quoting passages from the *Aeneid* and featuring Virgil himself as a central character.[105]

HESIOD *WORKS AND DAYS AND HOMERIC HYMNS*

Hesiod was an ancient Greek poet who wrote in roughly the same period as Homer. Composed around 700 BCE, the *Works and Days* is one of two major works attributed to Hesiod. *Works and Days* is primarily a moral text, with the aim of showing the reader how best to live the good life, a view that unifies what at first glance appears to be a miscellaneous collection of myths, maxims, and technical advice.[106] Hesiod prescribes a life of honest labor as the foundation of good living and attacks a life of idleness and usury.[107]

The *Homeric Hymns* is the name for a collection of thirty-three Greek poems composed in the epic style used by Homer. Once thought to be the work of Homer, due largely to their use of the Homeric meter (dactylic hexameter), they are now generally regarded as anonymous works composed across a long span of time from the time of Homer (7–6 BCE) through the Hellenistic period to Roman times.[108]

The hymns are directed toward ancient Greek gods, for example, *Hymn to Demeter*, and range in length from three to five hundred lines. The shortest of them are brief invocations, which are thought to have served as preludes to longer recitations of epic, including excerpts from Homer, at religious festivals.[109] The largest four hymns are complete epic narrative poems in their own right.

FOUR BRANCHES OF THE MABINOGI

The *Four Branches of the Mabinogi* or *Pedair Cainc Y Mabinogi* is a collection of Welsh stories preserved in two medieval manuscripts: the *White Book of Rhydderch* (*Llyfr Gwyn Rhydderch*) (1300–1325) and the *Red Book of Hergest* (*Llyfr Coch Hergest*) (1375–1425). Scholars regard the *Mabinogi* as part of a complex narrative tradition with their roots in earlier, oral storytelling.[110] The stories are the earliest known prose literature of Britain.[111]

The *Mabinogi* is divided into four main parts (or branches), which cover the events in the life and death of the legendary hero Pryderi, son of Pwyll. The stories are complex narratives comprised of mythology, romance, tragedy, and fantasy. Many of the stories contain instances of magical animals or of metamorphoses between the human and animal form (and vice versa). The story *Math Son of Mathonwy* (*Math fab Mathonwy*), which contains the tale of Bloddoweud, is the final in the four stories.

Lady Charlotte Guest published a bilingual translation, in collection with other Welsh stories, as *The Mabinogion* in 1877.[112] She is widely assumed to have coined the name *Mabinogion*, but the name first appears in 1795 in William Owen Pughe's translation.[113] Jones and Jones's 1949 translation pinpoints the name *mabynnogyon*, which appears once in the *Red Book of Hergest*, as a scribal error, an incorrect plural of Mabinogi.[114] The etymology for the name *Mabinogi* itself is unclear, but scholars regard it as relating to the stories of young people from the Welsh *Mab* meaning boy/son.[115]

OVID, *METAMORPHOSES*

Publius Ovid Naso (known to us as Ovid), was a Roman poet and contemporary of Virgil who lived and died during the reign of Emperor Augustus, the adopted son of Julius Caesar. His prior work *The Art of Love*, with themes of adultery and romantic conquest saw Ovid banished to Tomis on the Black Sea, a fort on the Roman frontier—a victim of Emperor Augustus's legislative program to reestablish Roman family values.[116] There he wrote two poems, including *Metamorphoses*, widely regarded to be his masterpiece.

Metamorphoses is a long narrative poem of twelve thousand lines divided into fifteen books.[117] For this work, Ovid abandoned the elegiac couplet used in his previous works and instead used the hexameter of Homeric epic.[118] The raw material for *Metamorphoses* is also of ancient Greece. The title is a Greek word meaning "changes of shape," the poem draws on ancient myths of Greek gods, which Ovid's audience worshipped under Roman names, and Ovid retells and often radically recasts a dizzying succession of more than 250 mythical transformations.[119] Human beings are turned into animals, plants, and stars, women are turned into men (and vice versa) and stones into human beings. Ovid also weaves in a description of creation based on the writings of the Stoic philosophers and ends with Roman myth and the wanderings of Aeneas.

Ovid's reworkings of the ancient Greek myths have been appropriated by artists and writers ever since their publication. Giants of European literature, including Chaucer, Boccaccio, Shakespeare, and Dante have all drawn heavily on the poem to create their own definitive works. Artists such as Titian and Gian Lorenzo Bernini have captured the shape shifting of antiquity in paint and marble. The extent of the artistic and cultural inspiration means that is not an exaggeration to regard *Metamorphoses* as one of the most influential works of all time.

NONNUS, *DIOYNSIACA*

Nonnus was an epic poet born in Hellenic Egypt, who is thought to have lived at Panopolis (now the city of Akhmim) sometime in the fifth century. Almost nothing else is known about his biography.

Dionysiaca is Nonnus's main work, an epic poem both in length and in style—20,426 lines in forty-eight books, composed in both the dialect and hexameter of Homer.[120] The principal narrative subject is the life of Dionysos, the god of the vine and winemaking, his travel to India, and his triumphant return, but the poem is of greater interest to students of mythology for the inclusion of all the stories of Greek mythology that Nonnus could find in earlier collections.[121]

Nonnus was part of an Alexandrian tradition that sought novelty in its treatment of the mythological material that had been the raw basis for stories for centuries. They sought out obscure variants and local myths, and a good number of the Greek myths, such as the fight between Dionysos and Perseus, are known only from Alexandrian writers such as Nonnus, or from even later compilers who drew on Alexandrian poetry.[122] This has led to some scholars' writing of the degenerate state of the Greek myths in *Dionysiaca*,[123] whereas others have started to recognize its intertextual value in the reading of earlier versions of the Greek myths.[124]

LIBRARY OF DIODORUS SICULUS

Diodorus Siculus (Diodorus of Sicily) was a Greek historian, born at Agyrium in Sicily in the first century BCE. These brief facts can be gleaned from Diodorus Siculus's own writing, but little else is known of his biography.

His major work *The Library of History*, is a vast compiled history of forty books arranged into three parts. The first part includes narratives of ancient mythologies from a number of discrete geographical regions, including Egypt and Greece. Some, such as the story of Phaethon and the Heliades are interwoven with an account of historical events, creating a kind of mythic history. Large parts of *The Library* have been lost, with only books 1–5 and 11–20 surviving.

THE *UPANIṢADS*

The *Upaniṣads* are a collection of sacred texts that comprise an extension of early Vedic thought known as Vedānta.[125] According to translator Patrick Olivelle, the *Upaniṣads* "represent some of the most important literary products in the history of Indian culture and religion" and, as foundational texts, have had a profound influence on the development of Hinduism.[126] The early *Upaniṣads* contain some of the most important philosophical concepts in Hinduism (and Jainism and Buddhism) such as the doctrine of rebirth, karma, and ascetic self-denial, but it is impossible to discern a unifying doctrine or philosophy across the (approximately) two hundred texts, composed across different regions and centuries.[127] The *Upaniṣads* also contain important cosmogonical and cosmological passages, such as the creation of the world from the body of the sacrificial horse in the *Brhadaranyaka Upaniṣad*,[128] and use myths to confer philosophical truths, for instance, the myth of Prajāpati leading the god-king Indra to the understanding of Ātman (the true self or "soul") as being simultaneously formless and ultimately real.[129]

Like the wider Vedic literature, the *Upaniṣads* were Brahmin-generated and -oriented texts "composed and at first transmitted from generation to generation orally and within their respective Vedic branches," a fact that is manifest in the text through the frequent use of deictic pronouns.[130] The dating and authorship of the *Upaniṣads* is uncertain. Some of the earliest and largest texts (e.g., the *Bṛhadaranyaka* and the *Chāndogya*) are compilations of material that is considered to have existed independently before their inclusion in these texts.[131] Olivelle considers that dating these texts with any more precision than a century "is as stable as a house of cards," but places the *Bṛhadaranyaka* and the *Chāndogya Upaniṣad* in the seventh to sixth centuries BCE "give or take a century so."[132]

THE TALE OF TWO BROTHERS

The Tale of the Two Brothers, is found in the Papyrus D'Orbiney (BM 10183) from the rule of Seti II (who ruled from 1200 to 1194 BC), the fifth ruler of the XIXth Dynasty. The narrative concerns the relationship between two brothers, the elder, married, brother, Anpu, and Bata, the younger brother who flees after his brother's wife attempts to seduce him. Bata mutilates his own genitals to demonstrate his fidelity to his brother (!), then goes to the Valley of the Cedar, where the story of the Persea tree takes place.

The *Tale of Two Brothers*, which came to light in 1852 with the discovery of the Papyrus D'orbiney, is one of the most studied of all the stories in the Egyptian Papyri and has been classified at various times as as myth, mythic folktale, or indeed "the oldest fairy tale in the world."[133] The excerpt is from Flinders-Petrie's *Egyptian Tales*, and is based almost entirely on the translation by Francis Llewellyn Griffith.[134]

FLINDERS RANGES DREAMING

The Adnyamathanha are an Aboriginal Australian people from the Flinders Ranges, the largest mountain range in South Australia.[135] Adnyamathanha were granted native title to much of their lands in 2009, but, like many Aboriginal peoples, they live predominantly in population centers outside of the Flinders Ranges.[136] The Adnyamathanha language, Yura Ngawarla, is still spoken by older people, but the language and many cultural traditions are in a state of decline.[137]

The story of the Iga tree, recorded by Dorothy Tunbridge in 1988, concerns the native orange (*Capparis mitchellii*) a favorite source of bush tucker.[138] The Iga is a small to medium-sized tree with dark green, oval-shaped leaves and showy white/cream flowers.[139] According to the myth, the trees were originally from Eromanga in Queensland and traveled down to the Flinders Ranges. Sometimes facing attack by other trees, such as the Coolibah (*Eucalyptus coolabah*), wherever they camped, Iga trees grew up.[140]

THE SPEAKING LAND

The Speaking Land is a collection of Australian Aboriginal oral literature from the Western Desert region and western Arnhem Land, Northern Territory. Told to the anthropologists Catherine Berndt and Ronald Berndt over careers spanning five decades, the 195 myths in this volume represent one of the largest collections of Australian Aboriginal mythology ever assembled. Numerous language groups are represented, but the Gunwinggu (Kunwinjku) dreaming stories are the most frequent—see Warramurrungundji.

HERBALS

John Gerard's *The Herball* or *Generall Historie* of Plants (1597) is one of the best known of the English herbals. A hugely influential text in its day, the first edition was the standard herbal for a generation until the publications of John Parkinson.[141] It is remarkable for its inclusion of a large number of woodcut figures (approximately 1,800 largely taken from early works), some of which are the first printed figures of important horticultural and agricultural plants, such as the potato.[142] *The Herball* is also remarkable for Gerard's relatively poor botanical knowledge (many of the woodblocks in the first edition are mismatched with the descriptions) and its less than scrupulous origins.

Gerard, a keen horticulturalist with a renowned garden at Holborn, had commissioned an English translation of Rembert Dodoen's (a Flemish physician and botanist) final work, the *Pemptades* (1583), but when the translator (a Dr. Priest) died before it was finished, Gerard completed the translation, altered the arrangement, and published it under his own name.[143] On top of this, Gerard's assertion that the legend of the Barnacle Tree was consistent with his own empirical observations has led to little respect for his work as a scientific text.[144] Notably, the historian of botany Agnes Arber contended that John Gerard "does not, it must be confessed, fully deserve the fame which has fallen to his share."[145]

Nicholas Culpeper (1616–1654) was the most famous exponent of what has come to be known as "astrological botany," a seventeenth-century approach to herbal medicine that linked both diseases and their botanical cures to celestial bodies or signs of the zodiac.[146] Arber approvingly cites the fierce criticisms of a contemporary, William Cole, but also notes that these rational criticisms did not stop the publication of edition after edition of Culpeper's works, including the *English Physitian* and the *Complete Herbal*.[147] In response to such criticism of Culpeper's credentials, some scholars highlight Culpeper as a key part of Puritan reforms aimed at making medical knowledge available to the poor and needy.[148]

TRAVELS OF JOHN MANDEVILLE

The Travels of John Mandeville (1357–1360) is a work of travel writing like no other. The book describes an epic journey of fabulous wanderings across Southwest Asia, North Africa, and India, yet many scholars doubt whether the author traveled at all. Indeed, there is much debate as to the authorship of *Mandeville's Travels*, with the list of candidates including "an English Knight, a Burgundian physician, a Liege liar and a Flemish monk."[149] Although the authorship is in dispute, the popularity of *Mandeville's Travels* is not—the work was one of the most popular secular texts of medieval times, and more than three hundred manuscripts survive to the present day.[150]

Sir Walter Raleigh is reported to have described Mandeville as the "greatest fabler of the world,"[151] and recent scholarship has attempted to portray Mandeville as "the world's greatest traveller."[152] However, not all scholars have been impressed by *Mandeville's Travels*. Thomas Browne, in his inimitable style, states baldly of the text that "it may afford commendable mythologie, but in a natural and proper Exposition, it containeth impossibilities and things inconsistent with the truth."[153]

NINE HERBS CHARM

The *Lacnunga*, a tenth-century Old English manuscript, sets down a cycle of pagan charms that collectively describe the virtues of nine herbs.[154] The best known of these is the *Nine Herbs Charm*, one which Braekman considers to reference only seven herbs,[155] but which does actually contain reference to nine plants.[156] The *Nine Herbs Charm* (indeed the entire *Lacnunga*) contains a noticeable Christian influence, with clear references to Christ, but it is often difficult to make a clear distinction between material of Christian origin, classical sources, and those which belong to Germanic or Celtic mythology.[157] The *Nine Herbs Charm* is one of only two Old English texts (the other being *Maxims I* of the Exeter Book) that contain any reference to Anglo-Saxon myths, with a short reference to Woden (Odin) taking nine twigs to cut an adder into nine parts.[158] Braekman notes that throughout the text "the herbs are regarded and addressed as powerful beings having a personality, a will, sometimes even a majesty of their own."[159]

KAKADU MAN

Big Bill Neidjie (c. 1920–2002) was an Aboriginal leader and the last surviving speaker of the Gagudju language, an Indigenous Australian language from northern Arnhem Land, Northern Territory.[160] Bill was born at Alawanydjawany on the East Alligator River, and like his father before him was a traditional owner of the Bunitj estate.[161] In the 1970s, Bill took the visionary and courageous step of leasing all of the Bunitj lands to the Commonwealth of Australia to aid the establishment of Kakadu National Park, now a World Heritage Area.[162] Not wishing to take customary secrets to the grave, as one of the few surviving Gadudju, Bill allowed recorded narratives to be collated into two books published during his lifetime, *Kakadu Man* (1985) and *Story About Feeling* (1989), in the hope of sharing his knowledge of sacred Aboriginal Law with the uninitiated. On one level, Bill's stories are descriptions of the surrounding natural world, reflections of Aboriginal history, and descriptions of cultural and social practices.[163] On another they are manifestations of a profoundly relational way of knowing about, and being in, the world—sprung from a deep knowledge of, and love for, country.

SHAHNAMA

The *Shahnama* (or *Book of Kings*) is a lengthy epic poem written by the Persian poet Firdausi (c. 1000 CE). The *Shahnama* combines myth with the early history of Persia (Iran) and at nearly sixty thousand couplets, is the longest epic poem ever written by a single author.[164] The material includes Sassanian cosmological accounts, heroic tales, and an account of the historical period covering the rule of the invading Aryan Keyumars to the Arab conquest of Persia and the fall of the Sassanid Empire in 651.[165] Much of this material involves a compilation of ancient histories and old mythical tales composed long before Firdausi's time.[166] The *Shahnama* is of great cultural significance in Persia (Iran) and ethnically linked regions such as Afghanistan.[167]

GREEK ALEXANDER ROMANCE

First recorded in the third century CE by Psuedo-Callisthenes, the Alexander Romance concerns the mythical adventures of Alexander the Great.[168] The Romance was translated into Latin in the fourth century and from there it was translated during the Middle Ages into every major language, spawning numerous versions.[169] These versions have had a significant role in spreading the Romance tales far and wide. Indeed, translator Richard Stoneman regards the Romance as being as influential on literature and art as the Gospels.[170] Even within the small niche of botanical mythology, this influence is borne out. A Syriac version had significant influence on Arabic literature, including tales of the *wak-wak* tree, and was a major source for Firdausi's *Shahnama*.[171]

ACARANGA SUTRA

The *Acaranga Sutra* (fifth–first century BCE) is the first of twelve angas, original texts of Jainism based on the teachings of the Mahavira, the twenty-fourth and last *tirthankara* (spiritual teacher) and a contemporary of the Buddha.[172] The *Acaranga Sutra* contains the most reliable story of the life of Mahavira as well as Mahavira's rules and conduct for ascetic life.[173] The text provides a detailed exploration of the nature of the living world and sets down a detailed moral code for human action toward living beings.[174]

Jainism remains an important religious tradition in India and is separated into two predominant sects, the Digambara and the Śvetāmbara. The religion is characterized by an extreme form of discipline, and adherents must, above all, practice *ahimsa*, or nonviolence toward living beings, from human beings to plants. The complex system of Jain ascetic rules set down in the *Acaranga Sutra* are directed to the liberation of the soul (*jiva*) from the endless cycle of rebirth.

EPIC OF GILGAMESH

The *Epic of Gilgamesh* is a narrative account of the adventures of Gilgamesh, King of Uruk (Warka in central Iraq), in his futile search for immortality. The classical *Epic* is an ancient Mesopotamian poem written in the cuneiform script of Akkadian, a broad linguistic group covering the Semitic Babylonian and Assyrian dialects spoken for two thousand years to the time of Alexander the Great's successors.[175] This is the longest and greatest of the Akkadian epics.[176] It is known that the *Epic was* composed by compiling existing Sumerian stories about Gilgamesh (or Bilgamesh as he is known in Sumerian), into a single work.[177] The stories are known only from fragments of clay tablets, meaning that their narrative and meaning is incomplete.[178] This history of composition has produced two principal versions, an Old Babylonian version dated to approximately the mid-third millennium BCE,[179] and the standard version, an Assyrian recension dating from the seventh century BC found on clay tablets in the ruins of the Royal Library of Ashurbanipal at Nineveh.[180]

Trade and cultural links across nations meant that the *Epic of Gilgamesh* was well known in antiquity and exerted an influence on the development of better-known stories from the region. The story of the flood clearly appears in the Old Testament, there are close similarities with tales in the *1001 Nights*, and Enkidu's (Gilgamesh's wild-man companion) visit to the underworld has parallels in Homer's *Odyssey*.[181]

HOMER, *ODYSSEY*

Broadly narrating the story of Odysseus's return from the Trojan War, the *Odyssey* (c. late eighth century BCE) is the later of two epic poems (the *Iliad* being the earlier) attributed to the ancient Greek poet Homer.[182] The *Odyssey* is one of the cornerstones of Western literature and has had an enormous (an ongoing) impact on Western culture—exemplified by the word *odyssey* in English being synonymous with epic voyage. The *Odyssey* is composed in Homer's dactylic hexameter, in the style of oral poetry, with some scholarly debate as to the extent to which Homer used writing to construct and compose his epics.[183] Regardless, much of the subject matter had been orally transmitted, from perhaps as far back as twelfth century BCE.[184] A mythological example of this oral base is provided by the story of Jason and the Argonauts. The *Odyssey* references an earlier, oral form of the myth, one of the first literary references before Appollonius's version some five hundred years later.[185]

NOTES

INTRODUCTION

1. Robert Cooper Seaton, trans., *Apollonius Rhodius: The Argonautica* (London: William Heinemann, 1912). *The Argonautica* sets down the old oral tale (to which Homer makes reference) of Jason and his epic journey to retrieve the Golden Fleece from the land of Colchis. The text of *The Argonautica* has itself been an inspiration for many chroniclers and re-tellers of myth, from the epic poets Ovid and Virgil to the animator Ray Harryhausen, whose cinematic presentation, like all good re-tellings of myth, differs slightly from prior versions. In *The Argonautica*, the teeth of the Colchian dragon, guardian of the Golden Fleece, are those sown by Aeëtes. These sown teeth produce a band of "earthborn people" who are the inspiration for the skeleton army. Although the outer forms vary slightly, the basic tale, of the terrifying warriors born from the teeth of a ferocious creature, remains in the movie. This fantastical metamorphosis utterly captivated my childhood imagination, although, in our backyard version, we used pebbles and sticks instead of dragon's teeth and swords.

2. William Doty, *Mythography: The Study of Myths and Rituals* (Tuscaloosa: University of Alabama Press, 2000), 6.

3. Ibid., 6–7.

4. Ibid.

5. Edward B. Tylor. *Primitive Culture.* (London: John Murray, 1920).

6. Doty, *Mythography*, 14.

7. Karen Armstrong, *A Short History of Myth* (Edinburgh: Canongate, 2006), 8–9.

8. Robert Segal, *Myth: A Very Short Introduction* (Oxford: Oxford University Press, 2004), 5.

9. Joseph Campbell, *The Hero with a Thousand Faces* (London: Fontana Press, 1993).

10. Ibid., 3.

11. Claude Lévi-Strauss, "The Structural Study of Myth," *The Journal of American Folklore* 68 (1955): 429.

12. Doty, *Mythography*, 80.

13. See Val Plumwood, *Feminism and the Mastery of Nature* (London and New York: Routledge, 1993).

14. See Val Plumwood, *Environmental Culture: The Ecological Crisis of Reason.* (London: Routledge, 2002).

15. Ibid., 8–9.

16. Val Plumwood, "Nature in the Active Voice," *Australian Humanities Review* 46 (2009), 113–29.

17. For the history of such instrumentalism in horticulture see chapter 1.

18. For a recent example of this see Annette Giesecke. *The Mythology of Plants: Botanical Lore from Ancient Greece and Rome* (Los Angeles: J. Paul Getty Museum, 2014).

19. Plumwood, *Environmental Culture*, 56.

20. For example, the classic text by Ernst Lehner and Johanna Lehner, *Folkore and Symbolism of Flowers, Plants and Trees* (Mineola, NY: Dover Publications, 2003).

21. Michael Marder, "The Life of Plants and the Limits of Empathy," *Dialogue* 51 (2012): 259–73.

22. Ibid., 267.

23. Plumwood, "Nature in the Active Voice," 126–27.

24. Plumwood, *Environmental Culture*, 56–57.

25. Plumwood, "Nature," 127.

26. Valentin N. Voloshinov, *Marxism and the Philosophy of Language* (Cambridge: Harvard University Press, 1986), 86.

27. Sean Hand, ed., *The Levinas Reader* (Oxford: Basil Blackwell, 1989), 43: "The relationship with the other is not an idyllic and harmonious relationship of communion, or a sympathy through which we put ourselves in the other's place; we recognize the other as resembling us, but exterior to us."

28. Marshall Sahlins, "What Kinship Is (Part One)," *Journal of the Royal Anthropological Institute* 17 (2011): 2–19.

29. Doty, *Mythography*, 72.

30. Plumwood, "Nature," 125.

31. Val Plumwood, *The Eye of the Crocodile*, ed. Lorraine Shannon (Canberra: ANU E-Press, 2012), 44.

32. Doty, *Mythography*, 83.

33. Ibid.

34. For this distinction, I am grateful to Sahlins, *Kinship*, 2.

35. Lévi-Strauss, "The Structural Study of Myth," 431.

36. Plumwood, "Nature," 127–28.

1. John Parkinson, "Epistle to the Reader," in *Paradisi in Sole Paradisus Terrestris* (London: Methuen, 1904), 24.

2. Ibid.

3. Gen. 1:26 ESV.

4. Matthew Hall, *Plants as Persons: A Philosophical Botany* (Albany: State University of New York Press, 2011). Such a view also blends biblical perspectives with those of Aristotle (see chapter 5).

5. Lynn White Jr., "The Historical Roots of Our Ecologic Crisis," *Science* 10 (1967): 1203–07.

6. Pope Francis, in his most recent encyclical (*Encyclical Letter Laudato Si Of the Holy Father Francis on Care for Our Common Home*), has stressed that although such anthropocentric interpretations of Genesis have previously prevailed, they are incorrect. Pope Francis draws on a range of Biblical material and more recent commentary to argue for the inherent value in a nonhuman nature which is itself a manifestation of God (s.67–69). He also draws upon the teachings of his namesake Saint Francis of Assisi to suggest bonds of kinship between human and nonhuman nature (s.87). Pope Francis does not, however, dispute the separation of human and nonhuman nature based upon the possession of intelligence and reason.

7. Ibid., 13.

8. Doty, *Mythography*, 5.

9. Ibid., 11.

10. Sahlins, *Kinship*, 2–19.

11. Ibid., 11.

12. Ibid.

13. Rupert Stasch, *Society of Others: Kinship and Mourning in a West Papuan Place* (Berkeley: University of California Press, 2009), 133.

14. After all, when researching your family tree, the fact of sharing a common ancestor is the one thing that demonstrates that people are your kin.

15. Arthur Gilchrist Brodeur, trans., *The Prose Edda by Snorri Sturluson* (Oxford: Oxford University Press, 1916), 20–22.

16. Bruce Lincoln. *Myth, Cosmos, and Society: Indo-European Themes of Creation and Destruction* (Cambridge: Harvard University Press, 1986).

17. F. Max Müller, trans., *The Upanishads Part 2* (Oxford: Clarendon Press, 1884), 73–74.

18. In China, the first being is not an ox (or a horse), but a tiger. According to Wu, a "well-known Yi creation epic in Yao'an, Yunan describes how the universe and all plants and things on the earth were gradually generated from the tiger's body." Wu, *Pangu Myth*, 368.

19. Edward William West, trans., *Pahlavi Texts Part 1* (Oxford: Clarendon Press, 1880), 53.

20. Edward Theodore Chalmers Werner, *Myths and Legends of China* (London: George G. Harrap, 1922), 77.

21. Ella Elizabeth Clark, *Indian Legends of the Pacific Northwest*, 2nd ed. (Berkeley and Los Angeles: University of California Press, 2003), 83.

22. John Martin Crawford, trans., *The Kalevala: The Epic Poem of Finland Vol. 1* (Cincinnati: Robert Clarke, 1898), 61.

23. Horace Hayman Wilson, trans., *The Vishnu Purana* (London: John Murray, 1840), 41.

24. Ernest Alfred Wallis Budge, trans., *The Egyptian Texts* (London: Kegan Paul, Trench and Trübner, 1912), 11–12.

25. See Salt Papyrus 10051.

26. W. F. Kirby, trans., *Kalevala: The Land of Heroes. Vol. 1* (London and Toronto: J. M. Dent and Sons, 1907), 10–11.

27. Delia Goetz and Sylvanus G. Morley, trans., *Popol Vuh: Sacred Book of the Ancient Quiche Maya from the Translation of Adrian Recinos* (Norman: University of Oklahoma Press, 1950) 165–67.

28. Matthew W. Stirling, "Origin Myth of Acoma and Other Records," *Smithsonian Institute, Bureau of American Ethnology Bulletin* 135 (1942): 1–2.

29. Matilda Coxe Stevenson, "Ethnobiology of the Zuni Indians," in *Thirteenth Annual Report of the Bureau of American Ethnology to the Secretary of the Smithsonian Institution 1908–1909*, 46.

30. Ibid.

31. William Edward Hanley Stanner, *White Man Got No Dreaming: Essays 1938–1973*. Canberra: ANU Press, 1979.

32. Deborah Rose, "A Social and Ecological Analysis of Dreaming Trees" (unpublished manuscript, 1987), 6–7.

33. John Bradley, *Yanyuwa Country: The Yanyuwa People of Borroloola Tell the History of Their Land* (Richmond, Victoria: Greenhouse, 1988), 12.

34. George Chaloupka. *Gundulk abel gundalg: Mayali flora* (Darwin, Australia: Northern Territory Museum of Arts and Sciences, 1984).

35. George Grey, *Polynesian Mythology* (London: John Murray, 1855), 1–15.

36. John White, *The Ancient History of the Maori, His Mythology and Traditions*

Volume 1 (Wellington: George Didsbury, Government Printer, 1887), 26–27.

37. William George Aston, trans., *Nihongi: Chronicles of Japan from the Earliest Times to A.D. 697* (London: Kegan Paul, Trench, Truebner, 1896), 58–59.

38. Ibid.

39. Lévi-Strauss, "The Structural Study of Myth," 430.

40. Sahlins, *Kinship*, 15.

41. White, *Maori History*, 142–43.

42. Elsdon Best, *The Māori, Volume 2* (Wellington: Harry H. Tombs, 1924), 452.

43. Deborah Rose, *Country of the Heart: An Indigenous Australian Homeland* (Canberra: Aboriginal Studies Press, 2002), 110.

44. John Bradley et al, *All Kind of Things from Country: Yanyuwa Ethnobiological Classification* (Brisbane: Aboriginal and Torres Strait Islander Studies Res Rep Series 6, 2006), 1.

45. Arthur Oncken Lovejoy. *The Great Chain of Being: A Study of the History of an Idea* (New York: Harper, 1936).

46. Christopher D. Stone, "Should Trees Have Standing? Towards Legal Rights for Natural Objects," *Southern California Law Review* 45 (1972): 450–501.

47. Gn 1:11–13, 26–30.

48. Brodeur, *Prose Edda*, 20–22.

49. Ralph Thomas Hotchkin Griffith, trans., *The Hymns of the Rig Veda Volume 4* (Benares: G. J. Lazarus, 1892), Book X.

50. West, *Pahlavi Texts*, 45–46.

51. Cited in Paul R. Goldin, "The Myth that China Has No Creation Myth," *Monumenta Serica* 56 (2008): 9.

52. Kirby, *Kalevala*, 10–11.

53. Goetz and Morley, *Popol Vuh*, 165–67.

54. Stirling, *Acoma Origin Myth*, 1–2.

55. Bradley, *Yanyuwa Country*, 12.

56. Chaloupka, *Mayali Flora*.

57. White, *Maori History*, 26–27.

58. Aston, *Nihongi*, 58–59.

CHAPTER 2. GODS

1. Maurice Platnauer, trans., *Claudian Vol. 1.* (Cambridge: Harvard University Press, 1922), 381.

2. Much of the discussion here is based on Judith M. Hadley. *The Cult of*

Asherah in Ancient Israel and Judah: Evidence for a Hebrew Goddess (Cambridge: Cambridge University Press, 2000).

3. See 2 Kgs 17:10; 18:4; 21:3; 23:13–14; Jer 17:2.

4. Is 1:29.

5. Matthew Hall, "Talk Among the Trees: Animist Plant Ontologies and Ethics," in *The Handbook of Contemporary Animism*, ed. Graham Harvey (Durham: Acumen Publishing, 2013), 385–94.

6. According to Folkard, "After Rome Pagan became Rome Christian, the priests of the Church of Christ recognised the importance of utilising the connexion which existed between plants and the old pagan worship, and bringing the floral world into active cooperation with the Christian Church by the institution of floral symbolism, which should be associated not only with the names of saints, but also with the Festivals of the Church." Richard Folkard, *Plant Legends, Lore, Lyrics* (London: Sampson Low, Marston, Searle, and Rivington, 1924), 40.

7. Mircea Eliade. *The Sacred and the Profane: The Nature of Religion* (New York: Harcourt, Brace, and World, 1959).

8. Müller, *Upanishads* Part 2, 21.

9. Benjamin Thorpe, trans., *The Poetic Edda* (Lapeer, MI: The Nothvegr Foundation Press, 2004), 6.

10. Brodeur, *Prose Edda,* 27.

11. Kirby, *Kalevala,* 12.

12. Ralph L. Roys, trans., *The Book of Chilam Balam of Chumayel* (Washington, DC: Carnegie Institution, 1933), 98–100.

13. Mircea Eliade, *Patterns in Comparative Religion* (New York: Sheed and Ward, 1958), 267.

14. David L. Haberman. *People Trees: Worship of Trees in Northern India* (Oxford: Oxford University Press, 2013).

15. Thorpe, *Poetic Edda,* 36.

16. Michael D. J. Bintley, "Plant Life in the Poetic Edda." In *Sensory Perception in the Medieval West,* ed. Simon Thomson and Michael Bintley (Turnhout, Belgium, Brepols, 2016), 235.

17. Ibid., 238.

18. Ibid., 242.

19. Haberman (2013: 15) points out this is common in broadly panentheistic "traditions that regard the world as an actual expression or manifestation of divinity."

20. *Padma Purana,* 2.1.15.

21. Haberman, *People Trees*, 78.

22. Ibid.

23. *Padma Purana* 2.1.15.

24. Swami Vijñanananda, trans., *The S'rîmad Devî Bhâgawatam*. 1921–22. Online at sacredtexts.com, 899.

25. Ibid.

26. E. A. Wallis Budge, trans., *The Book of the Dead: the Chapters of Coming Forth by Day* (London: Kegan Paul, Trench, Truebner, 1898), 109.

27. Ibid., 277.

28. Alfred Godley Denis, trans., *Herodotus Volume 1* (Cambridge: Harvard University Press, 1921), 342.

29. Bill Neidjie, *Story about Feeling* (New South Wales: Mybrood, 1985), 24.

30. Basil Hall Chamberlain, trans., *The Kojiki: Records of Ancient Matters* (Tokyo: Asiatic Society of Japan, 1919), 64–65.

31. Graham Harvey, *Animism: Respecting the Living World* (London: Hurst, 2005), 44.

32. Harris Rackham, trans., *Pliny: Natural History Volume 4* (London: William Heinemann, 1960), 547–51.

33. Ibid.

34. Alfred Denis Godley, trans., *Herodotus Volume 2* (Cambridge: Harvard University Press, 1921), 272–74.

35. Ralph Thomas Hotchkin Griffith, trans., *The Hymns of the Rig Veda Volume 3* (Benares: G. J. Lazarus, 1891), 265.

36. Gn 2:5–9.

37. André Ferdinand Herold, *The Life of Buddha*, trans. Paul C. Blum (New York: A. and C. Boni, 1927), 87–88.

38. Brodeur, *Prose Edda*, 27–30.

39. Kirby, *Kalevala*, 11–12.

40. Roys, *Chilam Balam*, 98–100.

41. *Padma Purana* 2.1.15.

42. Erik Pema Kunsang, *Dakini Teachings* (Boudhanath, Hong Kong, and Esby: Ranjung Yeshe Publications, 1999), xvii.

43. Harold North Fowler, trans., *Plato: Euthypro, Apology, Crito, Phaedo, Phaedrus.* (London: William Heinemann, 1943), 565.

44. Godley, *Herodotus*, 341–43.

45. Rackham, *Natural History Volume 4*, 547–51.

46. Chamberlain, *Kojiki*, 64-65.

47. Griffith, *Rig Veda*, 265–66.

1. John Healey, trans., *The City of God by Saint Augustine Volume 2* (Edinburgh: John Grant, 1909), 174.

2. Alexander Roberts and James Donaldson, eds., *The Ante-Nicene Fathers: Translations of The Writings of Our Fathers down to A.D. 325 Volume VIIII* (Buffalo: The Christian Literature Company, 1886), 199. Clement of Alexandria (c. 150–215 CE) is regarded as a founding father of the Christian church. Writing after Ovid, his proclamations against the metamorphoses were undoubtedly part of his wider mission to have Greeks abandon their pagan heritage in favor of Christianity.

3. Charles Martin, trans., *Ovid: Metamorphoses* (New York: Norton, 2004), 550.

4. Ibid., 526–27.

5. Steven Luper, "Annihilation," *Philosophical Quarterly* 37 (1987): 233–52.

6. See the myth of Adonis, where this creation of new life is made explicit.

7. Thorpe, *Poetic Edda*, 6.

8. Goetz and Morley, *Popol Vuh*, 84.

9. Ibid., 89.

10. Ibid.

11. Timothy Knowlton, *Maya Creation Myths: Words and Worlds of the Chilam Balam* (Boulder: University of Colorado Press, 2010), 126.

12. Hugh Gerard Evelyn-White, trans., *Hesiod, Homeric Hymns, and Homerica* (London, William Heinemann, 1914), 13.

13. Jennifer Larson, *Greek Nymphs: Myth, Cult, Lore* (Oxford: Oxford University Press, 2001), 29.

14. Clements Robert Markham, trans., *Narratives of the Rites and Laws of the Yncas* (New York: Burt Franklin, 1873), 4.

15. A highly resolved tree of life, based on completed genome sequences, shows the relatively close kinship of plants and human beings (green dot—*Oryza sativa*; red dot—*Homo sapiens*). http://itol.embl.de/itol.cgi. Source: Wikimedia Commons, Public Domain.

16. Robert Segal. *Theorizing About Myth* (Boston: University of Massachusetts Press, 1999).

17. Martin, *Metamorphoses*, 37.

18. Nonnus deviates from the tradition by telling of an Iris that sprung from Hyacinthos's blood. See William Henry Denham Rouse, trans., *Nonnus, Dionysiaca. Vol. 1* (Cambridge: Harvard University Press, 1940), 113.

19. Frank Justus Miller, trans., *Ovid: Metamorphoses Vol. 2* (London: William Heinemann, 1916), 79.

20. Alexander William Mair, trans., *Oppian, Colluthus, Tryphiodorus* (London: William Heinemann, 1928), 387.

21. Miller, *Metamorphoses Vol. 2*, 115–17. The life of Adonis demonstrates the two ends of the human-plant metamorphic cycle. Adonis was born from a tree and when he dies he once again becomes plant life. This plant life is fragile in turn. The very name anemone (from *anemos* = wind) illustrates how the winds of life can easily bring an end to beauty and vitality.

22. Charles Henry Oldfather, trans., *The Library of History of Diodorus Siculus. Vol. III* (Cambridge: Harvard University Press, 1939), 159–61.

23. Richard Nelson, *Make Prayers to Raven: A Koyukon View of the Northern Forest* (Chicago: University of Chicago Press, 1986).

24. Frank Justus Miller, trans., *Ovid: Metamorphoses Vol. 1* (London: William Heinemann, 1916), 455–57.

25. This punishment is echoed in Shakespeare's *The Tempest* with Ariel's imprisonment in the pine tree at the hands of the witch Sycorax.

26. Miller, *Metamorphoses Vol. 1*, 75.

27. Ibid., 161.

28. Indeed, in the myth of Myrrha and Adonis, both human death and birth are present. Myrrah was a princess who was transformed into a myrrh tree following incest with her father.

29. Edward Wilson Gifford, *Tongan Myths and Tales* (Honolulu: The Museum, 1924), 74.

30. Antony Alpers, *The World of the Polynesians* (Auckland: Oxford University Press, 1987), 271–72.

31. Robert D. Craig, *Handbook of Polynesian Mythology* (Santa Barbara: ABC CLIO, 2004), 160.

32. Patrick Olivelle, trans., *Upanisads* (London: Oxford University Press, 1996), 123.

33. Kisari Mohan Ganguli, *Mahabharata*, (Online at sacredtexts.com), 12.212, 95. The excerpt here refers to the fundamental concepts of karma and rebirth. Although the position of plants in the cycle of rebirth is often unclear (with plants often left out of many interpretations of Hindu rebirth), the *Mahābhārata* is very clear that all living creatures, including plants, participate in the revolving door of rebirth.

34. Olivelle, *Upanisads*, 38.

35. Ibid., 84.

36. Ibid., 142.

37. Deborah Rose, *Dingo Makes Us Human: Life and Land in an Aboriginal Australian Culture* (Cambridge: Cambridge University Press, 1992), 57.

38. Dorothy Tunbridge, *Flinders Ranges Dreaming* (Canberra: Aboriginal Studies Press, AIAS, 1988), 48.

39. Freya Mathews, *For Love of Matter: A Contemporary Panpsychism* (Albany: State University of New York Press, 2003), 79.

40. Ronald Murray Berndt and Catherine Helen Berndt, *The Speaking Land: Myth and Story in Aboriginal Australia* (Ringwood: Penguin Australia, 1989), 57–58.

41. Ibid., 62.

42. Ibid., 49.

43. Ibid., 50.

44. Nelson, *Make Prayers to Raven*, 16.

45. Ibid.

46. Marie Mauzé, "Northwest Coast Trees: From Metaphors for Culture to Symbols in Culture," in *The Social Life of Trees*, ed. Laura Rival (Oxford: Berg, 1998), 233–51. This creation of human beings from plants has strong parallels with the story in the *Völuspá* that humans were created from trees.

47. Goetz and Morley, *Popol Vuh*, 89.

48. Charlotte Guest, trans., *The Mabinogion* (London: Bernard Quaritch, 1877), 426.

49. Markham, *Narratives*, 4–5.

50. Miller, *Metamorphoses Vol. 2*, 99–101.

51. Frederick Cornwallis Conybeare, trans., *Philostratus: The Life of Apollonius of Tyana Vol. 1* (London, William Heinemann, 1912), 43–45.

52. Stephen Gaselee, trans., *The Love Romances of Parthenius and Other Fragments* (London, William Heinemann, 1916), 305–307.

53. Rouse, *Dionysiaca Vol. 1*, 111–13.

54. Miller, *Metamorphoses Vol. 2*, 79.

55. Ibid., 115–17.

56. Oldfather, *The Library of History of Diodorus Siculus*, 159–61.

57. Nelson, *Make Prayers to Raven*, 16.

58. Miller, *Metamorphoses Vol. 1*, 455–57.

59. Miller, *Metamorphoses Vol. 2*, 73–75.

60. Rouse, *Dionysiaca Vol. 3*, 461–67.

61. William Matthew Flinders Petrie, trans., *Egyptian Tales: Translated from the Papyri. Second Series XVIIIth to XIX Dynasty. Second Edition* (London: Methuen, 1913), 48.

62. Gifford, *Tongan Myths*, 72–74.

63. F. Max Müller, trans., *The Upanishads Part 1* (Oxford: Clarendon Press, 1879), 81–82.

64. Müller, *Upanishads Part 2*, 126–27.

65. Tunbridge, *Flinders Ranges Dreaming*, 62.

66. Berndt and Berndt, *Speaking Land*, 62.

67. Ibid., 57.

68. Ibid., 49.

CHAPTER 4. LEGEND

1. Harris Rackham, trans., *Pliny: Natural History Volume 4* (London: William Heinemann, 1960), 505.

2. Ibid.

3. Ibid.

4. Ibid., 77.

5. Ibid., 222.

6. Ibid.

7. John Gerard, *The herball, or Generall historie of plantes* (London: Printed by Adam Norton and Richard Whitakers, 1636), 1587–88.

8. John F. Healy, *Pliny the Elder on Science and Technology* (Oxford: Oxford University Press, 1999), 387.

9. Harris Rackham, trans., *Pliny: Natural History Volume 5* (London: William Heinemann, 1938), 169.

10. Hall, *Plants as Persons*, 38–41.

11. Rackham, *Natural History Vol. 4*, 505.

12. Ibid., 449.

13. John Mandeville, *The Travels of Sir John Mandeville; The Version of the Cotton Manuscript in Modern Spelling* (London: Macmillan, 1925), 174.

14. Ibid.

15. Folkard, *Plant Legends*, 118.

16. Robert Brown, trans., *Aeneas Silvius Piccolomini: Europe* (Washington, DC: The Catholic University of America Press, 2013), 32–33.

17. Ibid.

18. Ibid.

19. Quoted in Thomas Suarez, *Early Mapping of Southeast Asia: The Epic Story of Seafarers, Adventurers, and Cartographers Who First Mapped the Regions Between China and India* (North Clarendon, VT: Tuttle, 2013), 53.

20. Ibid.

21. Ibid.

22. Frances Jenkins Olcott, ed., *The Arabian Nights—Based on the Translation from the Arabic by Edward William Lane* (New York: Harold Holt, 1913), 1.

23. Quoted in Suarez, *Early Mapping*, 53.

24. Ibid.

25. Tess Anne Osbaldeston, trans., *Dioscorides: De Materia Medica* (Johannesburg: Ibidis Press, 2000), 624.

26. William Henry Samuel Jones, trans., *Pliny: Natural History Volume 7, Books 24–27* (London: William Heinemann, 1966), 242.

27. Eugene Willis Gudger, "Pliny's Historia naturalis: The Most Popular Natural History Ever Published," *Isis* 6 (1924): 270.

28. Ibid., 440.

29. Nicholas Culpeper, *Culpeper's English Physician and Complete Herbal* (London: Printed for the author, 1789), 251–52.

30. Folkard, *Plant Legends*, 109.

31. West, *Pahlavi Texts*, 31.

32. Ibid., 65.

33. Markham, *Narratives*, 124–25.

34. Ibid.

35. Richard Evans Schultes, Albert Hofmann, and Christian Rätsch, *Plants of the Gods: Their Sacred, Healing, and Hallucinogenic Powers* (Rochester, VT: Healing Arts Press, 1992), 140.

36. Godfrid Storms, *Anglo Saxon Magic* (s'Gravenhage: Martinus Nijhoff, 1948), 187.

37. Crawford, *Kalevala Vol. 1.*, 118–20.

38. Maurice Bloomfield, *Hymns of the Atharva Veda* (Oxford: Clarendon Press, 1897), 37. Hemp is, of course, well known, but the gangida is an unidentified tree of a similarly powerful nature. It is the subject of another line in the *Atharva Veda* which simply says "O gangida, a protector art thou, O gangida." Ibid., 38.

39. Ibid., 5.

40. Ibid., 42.

41. Monica Gagliano, Stefano Mancuso, and Daniel Robert, "Towards Understanding Plant Bioacoustics," *Trends in Plant Science* 17 (2012): 323–25.

42. Erazim Kohák, "Speaking to Trees ," *Critical Review* 6 (1992): 371–88.

43. Ibid.

44. Marilyn Strathern, *Kinship, Law, and the Unexpected: Relatives Are Always a Surprise.* (Cambridge: Cambridge University Press, 2005).

45. Gerard, *Herball*, 1587–88.

46. Mandeville, *Travels*, 174.

47. James Dallaway, *Constantinople Ancient and Modern, with Excursions to the Shores and Islands of the Archipelago and to the Troad* (London: T. Bensley, 1797), 395.

48. Olcott, *Arabian Nights*, 199.

49. John Bostock and Henry Thomas Riley, trans., *The Natural History of Pliny Volume 4* (London, Henry G Bohn, 1856), 218.

50. Culpeper, *Herbal*, 251–52.

51. West, *Pahlavi Texts*, 65.

52. Ibid., 30–31.

53. Markham, *Narratives*, 124–25.

54. Felix Grendon, "The Anglo Saxon Charms," *The Journal of American Folklore* 22 (1909): 191–95.

55. Crawford, *Kalevala Vol. 1*, 118–20.

56. Bloomfield, *Atharva Veda*, 41–42.

57. Ibid., 4–5.

CHAPTER 5. SENTIENCE

1. See Dawn Keetley and Angela Tenga, eds. *Plant Horror: Approaches to the Monstrous Vegetal in Fiction and Film* (London: Palgrave Macmillan, 2017).

2. Benjamin Jowett, trans., *The Dialogues of Plato: Volume 2* (New York: Charles Scribner's Sons, 1899), 568.

3. Ibid.

4. Ibid.

5. Hugh Lawson-Tancred, ed., *Aristotle: De Anima (On the Soul)* (Harmondsworth, Penguin, 1986).

6. William David Ross, ed., *Nicomachean Ethics* (Oxford: Clarendon Press, 1908). Online at https://ebooks.adelaide.edu.au/a/aristotle/nicomachean/.

7. Benjamin Jowett, trans., *The Politics of Aristotle Vol. 1* (Oxford: Clarendon Press, 1885), 14.

8. Paco Calvo, Paco, Vaidurya Sahi, and Anthony Trewavas "Are Plants Sentient?" *Plant, Cell & Environment* 40 (2017): 2858–69.

9. William Forbes Skene, trans., *The Four Ancient Books of Wales* (Edinburgh: Edmonston and Douglas, 1868), 281. While having nothing in common in terms of origins, the notion of fighting plants is similar to that found in the origin myth of the Iga tree, which the Adnyamathanha people tell in the Flinders Ranges of Australia (chapter 3).

10. Neidjie, *Kakadu Man,* 52.

11. This hadith is from the Sahih al-Bukhari, a widely respected collection of hadith compiled by Imam Muhammad al-Bukhari (d. 870 CE), containing more than 7,500 hadith in 97 books. M. Muhsin Khan, trans., *Sahih al-Bukhari* 918, Book 11, Hadith 42. Online at https://sunnah.com/bukhari/11/42.

12. Quoted in Salma Khadra Jayyusi, S. K. and Manuela Marin, eds., *The Legacy of Muslim Spain* (Leiden: Brill, 1992), 371.

13. Ibid.

14. Miller, *Metamorphoses Vol. 1,* 459.

15. Ganguli, *Mahabharata* 12.184, 26.

16. August Taber Murray, *Homer: The Iliad* (London: William Heinemann, 1928), 163.

17. Ibid., 277.

18. Ibid., 273.

19. Henry Rushton Fairclough, trans., *Virgil in Two Volumes. Vol. 1. Eclogues, Georgics. Aeneid I–VI.* (London: William Heinemann, 1916), 351.

20. Miller, *Metamorphoses Vol. 1,* 459.

21. Goetz and Morley, *Popol Vuh,* 75.

22. Fairclough, *Virgil,* 351.

23. Henry Wadsworth Longfellow, trans., *The Divine Comedy of Dante Alighieri* (Ludgate: George Routledge and Sons, 1867), 74–78.

24. William Forsell Kirby, trans., *Kalevala: The Land of Heroes Vol. 2* (London: J. M. Dent and Sons, 1907), 194–98.

25. Crawford, *Kalevala Vol. 1,* 205–206.

26. Arthur George Warner and Edmond Warner, trans., *The Shahnama of Firdausi Vol. 6* (London: Kegan Paul, Trench, Trübner, 1912), 167.

27. Ibid., 168.

28. Richard Stoneman, trans., *The Greek Alexander Romance* (London: Penguin Books, 1991), 134–35.

29. Ibid.

30. David Delgado Shorter, *We Will Dance Our Truth: Yaqui History in Yoeme Performances* (Lincoln: University of Nebraska Press, 2009), 137–38.

31. Ibid.

32. Johannes Adrianus Bernardus Van Buitenen, trans., *The Mahābhārata Vol. 1 Book 1* (Chicago: University of Chicago Press, 1973), 83.

33. James L. Fitzgerald, trans., *The Mahābhārata Vol. 7* (Chicago: University of Chicago Press, 2004), 564–67.

34. Ibid.

35. Larson, *Greek Nymphs*, 8.

36. Larson provides a very detailed taxonomy of Greek nymphs. See Ibid.

37. Larson, *Greek Nymphs,* 10.

38. Ibid. Nonnus's depiction of a speaking ash tree that reproaches Dionysos identifies the ash trees as a place which nymphs inhabit.

39. William H. S. Jones, trans., *Pausanias: Description of Greece Vol. 3* (London, William Heinemann, 1961).

40. Larson, *Greek Nymphs,* 33.

41. Evelyn-White, *Hesiod*, 425.

42. Alexander William Mair, trans.. *Callimachus and Lycophron* (London: William Heinemann, 1921), 91.

43. Doty, *Mythography*, 84.

44. Jgs 9:8–16.

45. Skene, *Four Ancient Books*, 276–81.

46. Neidjie, *Kakadu Man*, 52.

47. Miller, *Metamorphoses Vol. 1*, 457–61.

48. Ganguli, *Mahabharata*, 26.

49. Rushton Fairclough, *Virgil Vol. 1*, 349–53.

50. Crawford, *The Kalevala Vol. 1*, 205–206.

51. Kirby, *The Kalevala Vol. 2*, 194–98.

52. Warner and Warner, *Shahnama*, 167–68.

53. Stoneman, *Alexander Romance*, 134–35.

54. Ernest Alfred Wallace Budge, *The Life and Exploits of Alexander the Great: Being a Series of Translations of the Ethiopic Histories of Alexander by the Pseudo-Callisthenes and Other Writers* (London: C. J. Clay, 1896), 161–66.

55. Evelyn White, *Hesiod*, 425.

CHAPTER 6. VIOLENCE

1. See Gary L. Francione. *Animals as Persons* (New York: Columbia University Press, 2008).

2. Bhikku Bodhi, ed., *The Middle Length Discourses of the Buddha: A Translation of the Majjhima Nikaya.* 3rd ed. (Boston: Wisdom Publications, 2005), 169.

3. Hermann Jacobi, trans., *Gaina Sutras Part 1: The Âkârâṅga Sûtra, The Kalpa Sûtra* (Oxford: Clarendon Press, 1884), 80–81.

4. Hall, *Plants as Persons,* 85.

5. Michael Marder, "Is It Ethical to Eat Plants?" *Parallax* 19 (2013): 29–37.

6. Augustus Taber Murray, *Homer: The Odyssey Vol. 1* (Cambridge: Harvard University Press, 1945), 187.

7. Stirling, *Origin Myth of the Acoma*, 5–6.

8. Bloomfield, *Atharva Veda*, 189.

9. The description of the wrath of Demeter following Erysichthon's felling of the sacred oak is the closest that I am aware of.

10. Kirby, *Kalevala Vol. 1*, 14.

11. Ibid., 15–16.

12. Ibid., 16.

13. Wilson, *Vishnu Purana*, 110.

14. Kathryn Yusoff, "Aesthetics of Loss: Biodiversity, Banal Violence, and Biotic Subjects," *Transactions of the Institute of British Geographers* 37 (2012): 578–92.

15. Ibid., 580.

16. Ibid.

17. Ibid., 582.

18. Stirling, *Acoma Origin Myth*, 5–6.

19. Sahlins, *Kinship*, 2–3.

20. Anton Kozmin notes that the tale of Rātā is found throughout the Pacific in Tonga, Niue, Sämoa, Tuvalu, Tokelau, Pukapuka, Mangaia, Societies, Tuamotu, Mangareva, New Zealand, and Hawai'i. Anton Kozmin, "Tree of Rata: Towards a Reconstruction of a Proto-Polynesian Text," *The Journal of the Polynesian Society* 119 (2010): 393–400.

21. In Aotearoa/New Zealand, Rātā is the indigenous name for a number of endemic trees in the genus *Metrosideros*.

22. White, *Maori History*, 69–70.

23. Ibid.

24. Te Pakaka Tawhai quoted in Graham Harvey, *Animism*, 55.

25. Stephen Savage, "The Rarotongan Version of the Story of Rata," *Journal of the Polynesian Society* 19 (1910): 142–57, quoted in Kozmin, *Tree of Rata*, 394.

26. Ibid., 26.

27. Flinders Petrie, *Egyptian Tales*, 64–65.

28. Ibid.

29. Ibid.

30. Aston, *Nihongi*. 147.

31. Ibid.

32. Neidjie, *Story about Feeling*, 27.

33. Berndt and Berndt, *Speaking Land*, 62.

34. Ibid.

35. Wilson, *Vishnu Purana*, 110.

36. Harvey, *Animism*, 116.

37. A 2016 study has found that humans have destroyed 10 percent of the world's "wild" areas in the past twenty-five years. James Watson et al., "Catastrophic Declines in Wilderness Areas Undermine Global Environment Targets," *Current Biology* 26 (2016): 2929–34.

38. Calvo et al., "Are Plants Sentient"

39. Aston, *Nihongi*, 275.

40. Jacobi, *Gaina Sutras*, 80–81.

41. Murray, *Odyssey*, 187–89.

42. Stirling, *Acoma Origin Myth*, 5–6.

43. Bloomfield, *Atharva Veda*, 189.

44. Kirby, *Kalevala Vol. 1*, 158–60.

45. Wilson, *Vishnu Purana*, 110.

46. White, *Maori History*, 69–70.

47. Rakshasas are "flesh eating" or "man eating" beings of Hindu mythology, first described in the *Rig Veda*, Hymn 87 of the 10th Mandala.

48. Ganguli, *Mahabharata*, 335–36.

49. Ibid., 25–26.

50. Andrew George, *The Babylonian Gilgamesh Epic: Introduction, Critical Edition, and Cuneiform Texts, Volume 1 (Oxford: Oxford University Press, 2003)*, 46–47.

51. Flinders Petrie, *Egyptian Tales*, 62–65.

52. Aston, *Nihongi*, 147.

53. Berndt and Berndt, *Speaking Land*, 62.

EPILOGUE. IMAGINATION AND BEYOND

1. This has led some authors to question whether mythology exists in separation from its cultural context. See Sophie Heller. *The Absence of Myth* (Albany: State University of New York Press, 2005).

2. See Robert Segal, *Myth: A Very Short Introduction,* (Oxford: Oxford University Press, 2004), 76.

3. If a definition of ritual has been agreed on by scholars, I've yet to find it. For those inclined, William Doty provides an in-depth discussion of what ritual is in Doty, *Mythography*, 368–96.

4. Tom F. Driver, *The Magic of Ritual: Our Need for Liberating Rites That Transform Our Lives and Our Communities* (San Francisco: HarperCollins, 1991), 47.

5. Stone, *Should Trees Have Standing?*, 450–501.

6. Ronald Grimes, "Performance Is Currency in the Deep World's Gift Economy: An Incantory Riff for a Global Medicine Show," *Interdisciplinary Studies in Literature and Environment* 9, no. 1 (2002): 158.

7. Harvey, *Animism*, 55.

8. See Hall, *Plants as Persons,* 134.

9. Bill Neidjie, *Story about Feeling* (Broome, Australia: Magabala Books, 1998), 18.

10. Gagliano et al., "Plant Bioacoustics ," 323–25.

GUIDE TO THE TEXTS

1. Bruce K. Waltke, "The Creation Account in Genesis 1:1–3, Part IV: The Theology of Genesis 1," *Bibliotheca Sacra* 132 (1975): 327–42.

2. Ibid.

3. Robert Alter, *The Five Books of Moses* (New York: W. W. Norton, 2004), xii.

4. David M. Carr, *Reading the Fractures in Genesis* (Westminster: John Knox Press, 1996), 64.

5. Brodeur, *Prose Edda,* 13.

6. Ibid.

7. Terry Gunnell, "Eddic Poetry," in *A Companion to Old Norse-Icelandic Literature and Culture,* 2nd ed., ed. R. McTurk (Oxford: Blackwell, 2004), 82.

8. Ibid., 93–95.

9. Ibid.

10. Ibid., 94.

11. Ibid., 93–95.

12. Griffith, *Rig Veda.*

13. Michael Witzel, "The Development of the Vedic Canon and Its Schools: The Social and Political Milieu," *Harvard Oriental Series, Opera Minora* 2 (1997): 257–348.

14. Ibid., 260.

15. Frits Staal, *Discovering the Vedas: Origins, Mantras, Rituals, Insights* (London: Penguin, 2009), 107.

16. David Flattery and Martin Schwarz, *Haoma and Harmaline: The Botanical Identity of the Indo-Iranian Sacred Hallucinogen "Soma" and Its Legacy in Religion,*

Language, and Middle Eastern Folklore, in *Near Eastern Studies* 21 (Berkeley: University of California Press, 1989).

17. Lincoln. *Myth, Cosmos, and Society.*

18. David Leeming, *A Dictionary of Asian Mythology*, published online by Oxford University Press, 2001.

19. Ibid.

20. West, *Pahlavi Texts*, xxii.

21. Ibid., xxxii.

22. Ibid., xxiv.

23. Ibid., xxii.

24. Ibid., xxiii.

25. Wu Xiaodong, "The Rhinoceros Totem and Pangu Myth: An Exploration of the Archetype of Pangu," *Oral Tradition* 16 (2001): 364–80.

26. Paul R. Goldin. "The Myth that China Has No Creation Myth," *Monumenta Serica* 56 (2008): 1–22.

27. Wu, *Pangu Myth.*

28. Ibid., 369.

29. David E. Gay, "The Creation of the Kalevala, 1833–1849," *Jahrbuch für Volksliedforschung* 42 (1997): 63–77.

30. John B. Alphonso-Karkala, "Transformation of Folk Narratives into Epic Composition in Elias Lönnrot's Kalevala," *Jahrbuch für Volksliedforschung* 31 (1986): 13–28.

31. Urpo Vento, "The Role of the Kalevala in Finnish Culture and Politics," *Nordic Journal of African Studies* 1 (1992): 82–93.

32. Ibid.

33. Alf Hiltebeitel, *Rethinking The Mahabharata: A Reader's Guide to the Education of the Dharma King* (Chicago: University of Chicago Press, 2001), 1.

34. George Von Simson, *The Mythic Background of The Mahābhārata*, *IT* 12 (1984): 191–223.

35. Jarich G. Oosten. *The War of the Gods: The Social Code in Indo-European Mythology* (Abingdon, Oxon: Routledge, 2015).

36. Wilson, *Vishnu Purana*, vi.

37. Ibid.

38. Dominic Goodall, *Hindu Scriptures* (Delhi: Motilal Banarsidass, 1996), xxxviii.

39. Wilson, *Vishnu Purana.*

40. Ibid., v.

41. Ibid., lxvii.

42. Ludo Rocher, *The Puranas* (Wiesbaden: Otto Harrassowitz Verlag, 1996), 59–61.

43. Goetz and Morley, *Popol Vuh*, 17.

44. Ibid., 23.

45. John M. Woodruff, "Ma(r)king Popol Vuh," *Romance Notes* 51 (2011): 97–106.

46. Goetz and Morley, *Popol Vuh*, 5.

47. Ibid.

48. Ibid.

49. Ibid.

50. See Nathan C. Henne, "Untranslation: The Popol Wuj and Comparative Methodology," *CR: The New Centennial Review* 12 (2012): 107–49.

51. Goetz and Morley, *Popol Vuh*.

52. Roys, *Chilam Balam*.

53. Timothy Knowlton, *Maya Creation Myths: Words and Worlds of the Chilam Balam* (Boulder: University of Colorado Press, 2010), 2.

54. Eleanor Harrison-Buck, "Reevaluating Chronology and Historical Content in the Maya Books of Chilam Balam," *Ethnohistory* 61 (2014), 683.

55. Timothy Knowlton, "Dynamics of Indigenous Language Ideologies in the Colonial Redaction of a Yucatec Maya Cosmological Text," *Anthropological Linguistics* 50 (2008): 94; Dennis Tedlock, *Two Thousand Years of Mayan Literature* (Berkeley: University of California Press, 2010), 249.

56. Knowlton, *Maya Creation Myths*, 2.

57. Roys, *Chilam Balam*, 5–6.

58. Knowlton, *Maya Creation Myths*, 2–3.

59. Bernard A. Knapp and Sophia Antoniadou, "Archaeology, Politics, and the Cultural Heritage of Cyprus," in *Archaeology under Fire: Nationalism, Politics, and Heritage in the Eastern Mediterranean and Middle East*, ed. Lynn Meskell, 14 (London: Routledge, 1998).

60. Barry Pritzke, *A Native American Encyclopedia: History, Culture, and Peoples* (Oxford: Oxford University Press, 2001), 8.

61. Daryll Forde, "A Creation Myth from Acoma," *Folklore* 41 (1930), 370.

62. Ibid.

63. Kathryn A. Seton and John J. Bradley, "When You Have No Law You Are Nothing: Cane Toads, Social Consequences, and Management Issues." *The Asia Pacific Journal of Anthropology* 5 (2004): 208.

64. Ibid.

65. Bradley. *Yanyuwa Country*.

66. John J. Bradley, "The Social, Economic, and Historical Construction of Cycad Palms among the Yanyuwa," in *Social Archaeology of Australian Indigenous Societies*, ed. David, Bruno, Bryce Barker, and Ian J. McNiven, 161–81 (Canberra: Aboriginal Studies Press, 2006).

67. Ibid., 162.

68. Paul S. C. Tacon, "Identifying Sacred Landscapes in Australia: From Physical to Social," in *Contemporary Archaeology in Theory: The New Pragmatism*, ed, Robert W. Preucel and Stephen A. Mrozowski, 84 (Chichester: Blackwell, 2010).

69. Ibid.

70. George Chaloupka. *Gundulk abel gundalg: Mayali flora.* (Darwin, Australia: Northern Territory Museum of Arts and Sciences, 1984).

71. Nicholas Evans, Dunstan Brown, and Greville G. Corbett, "The Semantics of Gender in Mayali, Partially Parallel Systems and Formal Implementation," *Language* 78 (2002): 112.

72. Ibid.

73. Victoria University of Wellington Library. "Ancient History of the Māori." http://nzetc.victoria.ac.nz/tm/scholarly/tei-corpus-WaiAnci.html; accessed Feb. 6, 2017.

74. Michael *Reilly*, "John White: The Making of a Nineteenth-Century Writer and Collector of Maori Tradition," *New Zealand Journal of History* 23 (1989): 171.

75. Mark Derby, "Ossian in Aotearoa—'Ponga and Puhihuia' and the Re-Creation of Myth," *Journal of New Zealand Studies* 8 (2009): 33.

76. Michael Reilly, "Review of *Traditional Stories from Southern New Zealand: He Kōrero nō Te Wai Pounamu*, trans. and ed. Christine Tremewan," *New Zealand Journal of History* 38 (2004), 311.

77. Derby, "Ossian in Aotearoa," 29.

78. Ibid., 34.

79. Ibid., 37.

80. Aston, *Nihongi*, xiii.

81. Ibid.

82. Ibid., xii–xv.

83. Leeming, *Dictionary of Asian Mythology*.

84. Ibid.

85. Herold, *The Life of Buddha*.

86. Gudger, "Pliny's *Historia naturalis*," 270.

87. Ibid.

88. Ibid., 269.

89. Ibid., 269–70.

90. Ibid.

91. William E. Gwatkin, "Dodona, Odysseus, and Aeneas," *The Classical Journal* 57 (1961): 97–102.

92. Chamberlain, *Kojiki*, ii.

93. Isomae Jun'ichi and Sarah E. Thal, "Reappropriating the Japanese Myths: Motoori Norinaga and the Creation Myths of the Kojiki and Nihon shoki," *Japanese Journal of Religious Studies* 27 (2000): 17.

94. Coxe Stevenson, *Zuni*.

95. Ibid., 46.

96. Ibid.

97. Ibid.

98. Markham, *Narratives*, vii.

99. Ibid.

100. Ibid.

101. Evelyn-White, *Hesiod*, xix.

102. Ibid., xiv.

103. T. E. Page, trans. *The Aeneid of Virgil Books I–VI.* (London: Macmillan, 1967), viii.

104. Ibid.

105. Henry Wadsworth Longfellow trans., *The Divine Comedy of Dante Alighieri* (Ludgate: George Routledge and Sons, 1867), 74–78.

106. Evelyn-White, *Hesiod*, xix.

107. Ibid., 25.

108. Ibid., xxxiv–xxxix.

109. Ibid., xxxv.

110. Brynley Roberts, "Tales and Romances," in *A Guide to Welsh Literature, Volume One*, ed. A. O. H. Jarman and Gwilym Rees Hughes (Cardiff: University of Wales Press, 1992), 205.

111. Nikolai Tolstoy, *The Oldest British Prose Literature: The Compilation of the Four Branches of the Mabinogion.* Lampeter: Edwin Mellen, 2009.

112. Charlotte Guest, trans., *The Mabinogion.* London: Bernard Quaritch, 1877.

113. William Owen Pughe, "The Mabinogion, or Juvenile Amusements, Being Ancient Welsh Romances," *The Cambrian Register* (1795): 177–87.

114. Gywn Jones and Thomas Jones, trans., *The Mabinogion.* London: Everyman Library, 1949.

115. Patrick K. Ford, *The Mabinogi and Other Medieval Welsh Tales* (Oakland: University of California Press, 2008), 2.

116. Martin, *Metamorphoses*, x.

117. Ibid., xii.

118. Ibid.

119. Ibid., xiii.

120. William Henry Denham Rouse, trans., *Nonnus, Dionysiaca. Vol. 1* (Cambridge: Harvard University Press, 1940.

121. Ibid., xix.

122. Ibid., x–xii.

123. Ibid., x.

124. Neil Hopkinson, ed., *Studies in the Dionysiaca of Nonnus* (Cambridge: Cambridge Philological Society, 1994).

125. Leeming, *Dictionary of Asian Mythology.*

126. Olivelle, *The Early Upanisads*, 3–7.

127. Ibid.

128. Ibid., 37.

129. Leeming, *Dictionary of Asian Mythology.*

130. Olivelle, *The Early Upanisads*, 8.

131. Ibid., 11.

132. Ibid., 12.

133. See Alan Dundes, "Projective Inversion in the Ancient Egyptian 'Tale of Two Brothers,'" *The Journal of American Folklore* 115, no. 457/458 (2002): 378–94; and Sally D. Katary, "*The Two Brothers as Folktale: Constructing the Social Context,*" *The Journal of the Society for the Study of Egyptian Antiquities* 24 (1994): 39–70.

134. Flinders Petrie, *Egyptian Tales,* 4.

135. Bob Ellis, "Iga—The Tree That Walked," *South Australian Geographical Journal* 112 (2013): 23.

136. Ibid., 25.

137. Ibid.

138. Dorothy Tunbridge. *Flinders Ranges Dreaming* (Canberra: Aboriginal Studies Press, AIAS, 1988).

139. Ellis, *Iga*, 27.

140. Tunbridge, *Flinders Ranges Dreaming.*

141. Agnes Arber, *Herbals: Their Origin and Evolution. A Chapter in the History of Botany 1470–1670* (Cambridge: Cambridge University Press, 1953), 134.

142. Ibid., 130.

143. Ibid., 129.

144. Ibid., 130.

145. Ibid., 129.

146. Ibid., 261.

147. Ibid., 262–63.

148. Olav Thulesius, *Nicholas Culpeper: English Physician and Astrologer* (New York: St. Martin's Press, 1992).

149. Iain Macleod Higgins, ed., *The Book of John Mandeville: With Related Texts* (Indianapolis: Hackett, 2011), xix.

150. Leila, K. Norako, *The Travels of Sir John Mandeville*, http://d.lib.rochester.edu/crusades/text/the-travels-of-sir-john-mandeville; accessed Feb. 6, 2017.

151. I owe this observation to Francis Tobienne Jr., *Mandeville's Travails: Merging Travel, Theory, and Commentary* (Newark: University of Delaware Press, 2016), 28.

152. Giles Milton, *The Riddle and the Knight: In Search of John Mandeville, the World's Greatest Traveller* (New York: Picador, 2002).

153. Browne, *Pseudodoxia*, 22.

154. Storms, *Anglo Saxon Magic*, 16–24.

155. Willy L. Braekman notes that the "two remaining herbs, chervil and fennel, however, may be supplied from the charm that in the manuscript immediately follows, the Lay of the Nine Twigs of Woden." Willy L. Braekman, "Notes on Old English Charms," *Neophilologus* 64 (1980): 461–69.

156. Grendon, *Anglo Saxon Charms*, 191–95. These are mugwort, plantain, lamb's cress, "venom-loather," chamomile, nettle, crab-apple, chervil, and fennel.

157. Braekman, "Old English Charms," 463.

158. Karin Olsen, "The Lacnunga and Its Sources: The Nine Herbs Charm and *Wið Færstice* Reconsidered," *Revista canaria de estudios ingleses* 55 (2007): 23–32.

159. Braekman, "Old English Charms," 464.

160. Bill Neidjie, Stephen Davis, and Allan Fox, *Kakadu Man* (New South Wales: Mybrood, 1985), 26.

161. Ibid.

162. Ibid., 8.

163. Phillip Morrisey, "Bill Neidjie's *Story About Feeling*: Notes on its Themes and Philosophy," *Journal of the Association for the Study of Australian Literature* 15 (2015): 1.

164. Leeming, *Dictionary of Asian Mythology*.

165. Warner and Warner, *Shahnama*.

166. Ibid., 49–52.

167. Whole volumes have been written on this subject. See Charles Melville and Gabriel van den Berg, eds. *Shahnama Studies II: The Reception of Firdausi's Shahnama* (Leiden: Brill, 2012).

168. Albert Mugrdich Wolohojian, trans. *The Romance of Alexander the Great by Pseudo-Callisthenes* (New York: Columbia University Press, 1969).

169. Richard Stoneman, trans., *The Greek Alexander Romance* (Harmondsworth: Penguin, 1991), 7.

170. Ibid., 2.

171. Ibid., 2, 7.

172. Jacobi, *Gaina Sutras Part 1*, xlvii.

173. Robert E. Van Voorst, *Anthology of World Scriptures* (Boston: Cengage Learning, 2016), 113.

174. Jacobi, *Gaina Sutras Part 1*.

175. Stephanie Dalley, *Myths from Mesopotamia: Creation, the Flood, Gilgamesh, and Others* (Oxford: Oxford University Press, 2000), xv.

176. Ibid., 39.

177. Ibid., 42–44.

178. Ibid.

179. Andrew George, *The Babylonian Gilgamesh Epic: Introduction, Critical Edition, and Cuneiform Texts, Volume 1 (Oxford: Oxford University Press, 2003), 5.*

180. Dalley, *Myths from Mesopotamia*, 47.

181. Ibid., 47–48 and Andrew George, *Babylonian Gilgamesh, 70.*

182. E. V. Rieu and D. C. H. Rieu, trans., *Homer: The Odyssey* (London, Penguin, 2003), xi.

183. Ibid., xiv–xv.

184. Ibid., xv.

185. M. L. West, "Odyssey and Argonautica," *The Classical Quarterly* 55 (2005): 39–64.

BIBLIOGRAPHY

Alpers, Antony. *The World of the Polynesians*. Auckland: Oxford University Press, 1987.

Alphonso-Karkala, John. "Transformation of Folk Narratives into Epic Composition in Elias Lönnrot's Kalevala." *Jahrbuch für Volksliedforschung* 31 (1986): 13–28.

Alter, Robert. *The Five Books of Moses*. New York: W. W. Norton, 2004.

Arber, Agnes. *Herbals: Their Origin and Evolution, A Chapter in the History of Botany 1470–1670*. Cambridge: Cambridge University Press, 1953.

Armstrong, Karen. *A Short History of Myth*. Edinburgh: Canongate, 2006.

Aston, William George, trans. *Nihongi: Chronicles of Japan from the Earliest Times to A.D. 697*. London: Kegan Paul, Trench, Truebner, 1896.

Bauer, Brian S., Vania Smith-Oka, and Gabriel E. Cantarutti, trans. *Account of the Fables and Rites of the Incas*. Austin: University of Texas Press, 2011.

Berndt, Ronald Murray, and Catherine Helen Berndt. *The Speaking Land: Myth and Story in Aboriginal Australia*. Ringwood: Penguin Australia, 1989.

Best, Elsdon. *The Maori, Volume 2*. Wellington: Harry H. Tombs, 1924.

The Bible, English Standard Version. biblegateway.com; accessed February 14, 2016.

Bintley, Michael D. J. "Plant Life in the Poetic Edda." In *Sensory Perception in the Medieval West*, edited by Simon Thomson and Michael Bintley. Turnhout, Belgium: Brepols, 2016.

Bloomfield, Maurice. *Hymns of the Atharva Veda*. Oxford: Clarendon Press, 1897.

Bodhi, Bhikku, ed. *The Middle Length Discourses of the Buddha: A Translation of the Majjhima Nikaya*. 3rd ed. Boston: Wisdom Publications, 2005.

Bostock, John, and Henry Thomas Riley, trans. *The Natural History of Pliny Volume 4*. London, Henry G Bohn, 1856.

Bradley, John. "The Social, Economic, and Historical Construction of Cycad Palms among the Yanyuwa." In *Social Archaeology of Australian Indigenous Societies*, edited by Bruno David, Bryce Barker, and Ian J. McNiven, 161–81. Canberra: Aboriginal Studies Press, 2006.

———. *Yanyuwa Country: The Yanyuwa People of Borroloola Tell the History of Their Land*. Richmond, Victoria: Greenhouse Publications, 1988.

———, et al. *All Kind of Things from Country: Yanyuwa Ethnobiological Classification*. Brisbane: Aboriginal and Torres Strait Islander Studies Res Rep Series 6, 2006.

Braekman, Willy L. "Notes on Old English Charms," *Neophilologus* 64 (1980): 461–69.

Brodeur, Arthur Gilchrist, trans. *The Prose Edda by Snorri Sturluson*. Oxford: Oxford University Press, 1916.

Brown, Robert, trans. *Aeneas Silvius Piccolomini: Europe*. Washington, DC: The Catholic University of America Press, 2013.

Browne, Thomas. *Pseudodoxia Epidemica* or, *Enquiries into Very Many Received Tenents, and Commonly Presumed Truths*. Printed by R. W. for N. Ekins, at the Gun in Paul's church-yard, 1658.

Budge, Ernest Alfred Wallace. *The Life and Exploits of Alexander the Great: Being a Series of Translations of the Ethiopic Histories of Alexander by the Pseudo-Callisthenes and Other Writers*. London: C. J. Clay, 1896.

Calvo, Paco, Vaidurya Sahi, and Anthony Trewavas. "Could Plants Be Sentient?" Pre-print uploaded to bioRxiv https://doi.org/10.1101/121731.

Campbell, Joseph. *The Hero with a Thousand Faces*. London: Fontana Press, 1993.

Carr, David. *Reading the Fractures in Genesis*. Westminster: John Knox Press, 1996.

Chaloupka, George. *Gundulk abel gundalg: Mayali flora*. Darwin, Australia: Northern Territory Museum of Arts and Sciences, 1984.

Chamberlain, Basil Hall, trans. *The Kojiki: Records of Ancient Matters*. Tokyo: Asiatic Society of Japan, 1919.

Clark, Ella Elizabeth. *Indian Legends of the Pacific Northwest*. 2nd ed. Berkeley and Los Angeles: University of California Press, 2003.

Clements, Robert Markham, trans. *Narratives of the Rites and Laws of the Yncas*. New York: Burt Franklin, 1873.

Conybeare, Frederick Cornwallis, trans. *Philostratus: The Life of Apollonius of Tyana Vol. 1*: London, William Heinemann, 1912.

Coxe Stevenson, Matilda. "Ethnobiology of the Zuni Indians." In *Thirteenth Annual Report of the Bureau of American Ethnology to the Secretary of the Smithsonian Institution 1908–1909*. Washington, DC.

Craig, Robert. *Handbook of Pacific Mythology*. Santa Barbara: ABC CLIO, 2004.

Crawford, John Martin, trans. *The Kalevala: The Epic Poem of Finland Vol. 1*. Cincinnati: Robert Clarke, 1898.

Culpeper, Nicholas. *Culpeper's English Physician and Complete Herbal*. London: Printed for the author, 1789.

Dallaway, James. *Constantinople Ancient and Modern, with Excursions to the Shores and Islands of the Archipelago and to the Troad*. London: T. Bensley, 1797.

Dalley, Stephanie. *Myths from Mesopotamia: Creation, the Flood, Gilgamesh, and Others*. Oxford: Oxford University Press, 2000.

Derby, Mark. "Ossian in Aotearoa—'Ponga and Puhihuia' and the Re-Creation of Myth." *Journal of New Zealand Studies* 8 (2009): 29–40.

Deshpande, N. A., trans. *The Padma Purana in Ten Volumes*, Delhi: Motilal Banarsidass, 1951.

Doty, William. *Mythography: The Study of Myths and Rituals*. Tuscaloosa: University of Alabama Press, 2000.

Driver, Tom. *The Magic of Ritual: Our Need for Liberating Rites That Transform Our Lives and Our Communities*. San Francisco: HarperCollins, 1991.

Dundes, Alan. "Projective Inversion in the Ancient Egyptian 'Tale of Two Brothers.'" *The Journal of American Folklore* 115, no. 457/458 (2002): 378–94.

Eliade, Mircea. *Patterns in Comparative Religion*. New York: Sheed and Ward, 1958.

———. *The Sacred and the Profane: The Nature of Religion*. New York: Harcourt, Brace and World, 1959.

Ellis, Bob. "Iga—The Tree That Walked." *South Australian Geographical Journal* 112 (2013): 23–36.

Evans, Nicholas, Dunstan Brown, and Greville G. Corbett. "The Semantics of Gender in Mayali, Partially Parallel Systems and Formal Implementation." *Language* 78 (2002): 111–55.

Evelyn-White, Hugh Gerard, trans. *Hesiod, Homeric Hymns, and Homerica*. London, William Heinemann, 1914.

Fairclough, Henry Rushton, trans. *Virgil in Two Volumes. Vol 1. Eclogues, Georgics. Aeneid I-VI*. London: William Heinemann, 1916.

Fairclough, Henry Rushton, trans. *Virgil in Two Volumes. Vol. 2. Aeneid VII–XII, The Minor Poems*. London: William Heinemann, 1934.

Fitzgerald, James, trans. *The Mahābhārata Vol. 7*. Chicago: University of Chicago Press, 2004.

Flattery, David, and Martin Schwarz. *Haoma and Harmaline: The Botanical Identity of the Indo-Iranian Sacred Hallucinogen "Soma" and Its Legacy in Religion, Language, and Middle Eastern Folklore*. Berkeley: University of California Press, 1989.

Flinders Petrie, William Matthew, trans. *Egyptian Tales: Translated from the Papyri. Second Series XVIIIth to XIX Dynasty.* 2nd Ed. London: Methuen, 1913.

Folkard, Richard. *Plant Legends, Lore, Lyrics.* London: Sampson Low, Marston, Searle, and Rivington, 1924.

Ford, Patrick K. *The Mabinogi and Other Medieval Welsh Tales.* Oakland: University of California Press, 2008.

Forde, Daryll. "A Creation Myth from Acoma." *Folklore* 41 (1930): 359–87.

Fowler, Harold North, trans. *Plato: Euthypro, Apology, Crito, Phaedo, Phaedrus.* London: William Heinemann, 1943.

Francione, Gary. *Animals as Persons.* New York: Columbia University Press, 2008.

Gagliano, Monica, Stefano Mancuso, and Daniel Robert. "Towards Understanding Plant Bioacoustics." *Trends in Plant Science* 17 (2012): 323–25.

Ganguli, Kisari Mohan. *Mahabharata.* Online at sacredtexts.com. Accessed November 4, 2016.

Gaselee, Stephen, trans. *The Love Romances of Parthenius and Other Fragments.* London: William Heinemann, 1916.

Gay, David. "The Creation of the Kalevala, 1833–1849." *Jahrbuch für Volksliedforschung* 42 (1997): 63–77.

George, Andrew. *The Epic of Gilgamesh.* London: Penguin Books, 1999.

———. *The Babylonian Gilgamesh Epic: Introduction, Critical Edition, and Cuneiform Texts, Volume 1.* Oxford: Oxford University Press, 2003.

Gerard, John. *The Herball, or Generall historie of plantes.* London: Printed by Adam Norton and Richard Whitakers, 1636.

Giesecke. Annette. *The Mythology of Plants: Botanical Lore from Ancient Greece and Rome.* Los Angeles: J. Paul Getty Museum, 2014.

Gifford, Edward Wilson. *Tongan Myths and Tales.* Honolulu: The Museum, 1924.

Godley, Alfred Denis, trans. *Herodotus Volumes 1 & 2.* Cambridge: Harvard University Press, 1921.

Goetz, Delia, and Sylvanus G. Morley, trans. *Popol Vuh: Sacred Book of the Ancient Quiche Maya from the Translation of Adrian Recinos.* Norman: University of Oklahoma Press, 1950.

Goldin, Paul R. "The Myth that China Has No Creation Myth." *Monumenta Serica* 56 (2008): 1–22.

Goodall, Dominic. *Hindu Scriptures.* Delhi: Motilal Banarsidass, 1996.

Grendon, Felix. "*The Anglo Saxon Charms.*" *The Journal of American Folklore* 22 (1909), 105–237.

Grey, George. *Polynesian Mythology*. London: John Murray, 1855.

Griffith, Ralph Thomas Hotchkin, trans. *The Hymns of the Rig Veda Volume 3*. Benares: G. J. Lazarus, 1891.

———, trans. *The Hymns of the Rig Veda Volume 4*. Benares: G. J. Lazarus, 1892.

Grimes, Ronald. "Performance Is Currency in the Deep World's Gift Economy: An Incantory Riff for a Global Medicine Show." *Interdisciplinary Studies in Literature and Environment* 9, no. 1 (2002): 149–64.

Gudger, Eugene Willis. "Pliny's *Historia naturalis*: The Most Popular Natural History Ever Published." *Isis* 6 (1924): 269–81.

Guest, Charlotte, trans. *The Mabinogion*. London: Bernard Quaritch, 1877.

Gunnell, Terry. "Eddic Poetry." In *A Companion to Old Norse-Icelandic Literature and Culture, Second Edition*, edited by Rory McTurk. Oxford: Blackwell, 2004.

Gwatkin, William E. "Dodona, Odysseus, and Aeneas." *The Classical Journal* 57 (1961): 97–102.

Haberman, David. *People Trees: Worship of Trees in Northern India*. Oxford: Oxford University Press, 2013.

Hadley, Judith. *The Cult of Asherah in Ancient Israel and Judah: Evidence for a Hebrew Goddess*. Cambridge: Cambridge University Press, 2000.

Hall, Matthew. *Plants as Persons: A Philosophical Botany*. Albany: State University of New York Press, 2011.

———. "Talk Among the Trees: Animist Plant Ontologies and Ethics." In *The Handbook of Contemporary Animism*, edited by Graham Harvey, 385–94. Durham: Acumen, 2013.

Hand, Sean, ed. *The Levinas Reader*. Oxford: Basil Blackwell, 1989.

Harrison-Buck, Eleanor. "Reevaluating Chronology and Historical Content in the Maya Books of Chilam Balam." *Ethnohistory* 61 (2014): 681–713.

Harvey, Graham. *Animism: Respecting the Living World*. London: Hurst, 2005.

Healey, John, trans. *The City of God by Saint Augustine Volume 2*. Edinburgh: John Grant, 1909.

Healy, John F., trans. *Pliny the Elder on Science and Technology*. Oxford: Oxford University Press, 1999.

Heller, Sophie. *The Absence of Myth*. Albany: State University of New York Press, 2005.

Henne, Nathan. "Untranslation: The Popol Wuj and Comparative Methodology." *CR: The New Centennial Review* 12 (2012): 107–49.

Herold, Andre Ferdinand. *The Life of Buddha*. Translated by Paul C. Blum. New York: A. and C. Boni, 1927.

Higgins, Iain Macleod, ed. *The Book of John Mandeville: With Related Texts.* Indianapolis: Hackett, 2011.

Hiltebeitel, Alf. *Rethinking The Mahabharata: A Reader's Guide to the Education of the Dharma King.* Chicago: University of Chicago Press, 2001.

Hopkinson, Neil, ed. *Studies in the Dionysiaca of Nonnus.* Cambridge: Cambridge Philological Society, 1994.

Jacobi, Hermann, trans. *Gaina Sutras Part 1: The Âkârânga Sûtra, The Kalpa Sûtra.* Oxford: Clarendon Press, 1884.

Jayyusi, Salma Khadra, and Manuela Marin, eds. *The Legacy of Muslim Spain.* Leiden: Brill, 1992.

Jones, Gywn, and Thomas Jones, trans. *The Mabinogion.* London: Everyman's Library, 1949.

Jones, William Henry Samuel, trans. *Pausanias: Description of Greece Vol. 3.* London: William Heinemann, 1961.

———, trans. *Pliny: Natural History Volume 7, Books 24–27.* London: William Heinemann, 1966.

Jowett, Benjamin, trans. *The Dialogues of Plato: Volume 2.* New York: Charles Scribner's Sons, 1899.

———, trans. *The Politics of Aristotle Vol 1.* Oxford: Clarendon Press, 1885.

Jun'ichi, Isomae, and Sarah E. Thal. "Reappropriating the Japanese Myths: Motoori Norinaga and the Creation Myths of the Kojiki and Nihon shoki." *Japanese Journal of Religious Studies* 27 (2000): 15–39.

Katary, Sally D. "*The Two Brothers as Folktale: Constructing the Social Context.*" *The Journal of the Society for the Study of Egyptian Antiquities* 24 (1994): 39–70.

Keetley, Dawn, and Angela Tenga, eds. *Plant Horror: Approaches to the Monstrous Vegetal in Fiction and Film.* London: Palgrave Macmillan, 2017.

Khan, M. Muhsin. trans. *Sahih al-Bukhari* 918, Book 11, Hadith 42. https://sunnah.com/bukhari/11/42; accessed March 28, 2016.

Kirby, William Forsell, trans. *Kalevala: The Land of Heroes. Vol. 1.* London and Toronto: J. M. Dent and Sons, 1907.

———, trans. *Kalevala: The Land of Heroes Vol. 2.* London: J. M. Dent and Sons, 1907.

Knapp, Bernard A., and Sophia Antoniadou. "Archaeology, Politics, and the Cultural Heritage of Cyprus." In *Archaeology under Fire: Nationalism, Politics, and Heritage in the Eastern Mediterranean and Middle East*, edited by Lynn Meskell, 13–43. London: Routledge, 1998.

Knowlton, Timothy. "Dynamics of Indigenous Language Ideologies in the

Colonial Redaction of a Yucatec Maya Cosmological Text." *Anthropological Linguistics* 50 (2008): 90–112.

———. *Maya Creation Myths: Words and Worlds of the Chilam Balam*. Boulder: University of Colorado Press, 2010.

Kohák, Erazim. "Speaking to Trees." *Critical Review* 6 (1992): 371–88.

Kozmin, Anton. "Tree of Rata: Towards a Reconstruction of a Proto-Polynesian Text." *The Journal of the Polynesian Society* 119 (2010): 393–400.

Kunsang, Erik Pema. *Dakini Teachings*. Boudhanath, Hong Kong, and Esby: Ranjung Yeshe Publications, 1999.

LaFleur, William. "Saigyō and the Buddhist Value of Nature." In *Nature in Asian Traditions of Thought: Essays in Environmental Philosophy,* edited by J. Baird Callicott and Roger T. Ames, 183–212. Albany: State University of New York Press, 1989.

Larson, Jennifer. *Greek Nymphs: Myth, Cult, Lore*. Oxford: Oxford University Press, 2001.

Lawson-Tancred, Hugh, ed. *Aristotle: De Anima (On the Soul)*. Harmondsworth: Penguin, 1986.

Leeming, David. *A Dictionary of Asian Mythology*, Published online by Oxford University Press, 2001.

Lehner, Ernst, and Johanna Lehner. *Folkore and Symbolism of Flowers, Plants, and Trees*. Mineola, NY: Dover, 2003.

Lévi-Strauss, Claude. "The Structural Study of Myth." *The Journal of American Folklore* 68 (1955): 428–44.

Lincoln, Bruce. *Myth, Cosmos, and Society: Indo-European Themes of Creation and Destruction*. Cambridge: Harvard University Press, 1986.

Longfellow, Henry Wadsworth, trans. *The Divine Comedy of Dante Alighieri*. Ludgate: George Routledge and Sons, 1867.

Lovejoy, Arthur Oncken. *The Great Chain of Being: A Study of the History of an Idea*. New York: Harper, 1936.

Luper, Steven. "Annihilation." *Philosophical Quarterly* 37 (1987): 233–52.

Mair, Alexander William, trans. *Callimachus and Lycophron*. London: William Heinemann, 1921.

———, trans., *Oppian, Colluthus, Tryphiodorus*. London: William Heinemann, 1928.

Mandeville, John. *The travels of Sir John Mandeville; The Version of the Cotton Manuscript in Modern Spelling*. London: Macmillan, 1925.

Marder, Michael. "Is It Ethical to Eat Plants?" *Parallax* 19 (2013): 29–37.

———. "The Life of Plants and the Limits of Empathy." *Dialogue* 51 (2012): 259–73.

Markham, Clements Robert, trans., *Narratives of the Rites and Laws of the Yncas*. New York: Burt Franklin, 1873.

Martin, Charles, trans. *Ovid: Metamorphoses*. New York: Norton, 2004.

Mathews, Freya. *For Love of Matter: A Contemporary Panpsychism*. Albany: State University of New York Press, 2003.

Mauzé, Marie. "Northwest Coast Trees: From Metaphors for Culture to Symbols in Culture." In *The Social Life of Trees*, edited by Laura Rival, 233–51. Oxford: Berg, 1998.

Melville, Charles, and Gabriel van den Berg, eds. *Shahnama Studies II: The Reception of Firdausi's Shahnama*. Leiden: Brill, 2012.

Miller, Frank Justus, trans. *Ovid: Metamorphoses Vol. 1*. London: William Heinemann, 1916.

———, trans. *Ovid: Metamorphoses Vol. 2*. London: William Heinemann, 1916.

Milton, Giles. *The Riddle and the Knight: In Search of John Mandeville, the World's Greatest Traveller*. New York: Picador, 2002.

Morrisey, Phillip. "Bill Neidjie's *Story About Feeling*: Notes on Its Themes and Philosophy." *Journal of the Association for the Study of Australian Literature* 15 (2015): 1–11.

Müller, F. Max, trans. *The Upanishads Part* 1. Oxford: Clarendon Press, 1879.

———, trans. *The Upanishads Part* 2. Oxford: Clarendon Press, 1884.

Murray, Augusts Taber. *Homer: The Iliad*. London: William Heinemann, 1928.

———. *Homer: The Odyssey Vol. 1*. Cambridge: Harvard University Press, 1945.

Neidjie, Bill. *Story About Feeling*. Broome, Australia: Magabala Books, 1998.

———, Stephen Davis, and Allan Fox. *Kakadu Man*. New South Wales: Mybrood, 1985.

Nelson, Richard K. *Make Prayers to Raven: A Koyukon View of the Northern Forest*. Chicago: University of Chicago Press, 1986.

Norako, Leila, K. "The Travels of Sir John Mandeville." http://d.lib.rochester.edu/crusades/text/the-travels-of-sir-john-mandeville; accessed March 12, 2017.

Olcott, Frances Jenkins, ed. *The Arabian Nights—Based on the Translation from the Arabic by Edward William Lane*. New York: Harold Holt, 1913.

Oldfather, Charles Henry, trans. *The Library of History of Diodorus Siculus. Vol. III*. Cambridge: Harvard University Press, 1939.

Olivelle, Patrick, trans. *The Early Upanisads*. Oxford: Oxford University Press, 1998.

———, trans. *Upanisads*. London: Oxford University Press, 1996.

Olsen, Karin. "The Lacnunga and Its Sources: The Nine Herbs Charm and *Wið Færstice* Reconsidered." *Revista canaria de estudios ingleses* 55 (2007): 23–32.

Oosten, Jarich. *The War of the Gods: The Social Code in Indo-European Mythology.* Abingdon: Routledge, 2015.

Osbaldeston, Tess Anne, trans. *Dioscorides: De Materia Medica.* Johannesburg: Ibidis Press, 2000.

Page, Thomas Ethelbert, trans. *The Aeneid of Virgil Books I–VI.* London: Macmillan, 1967.

Parkinson, John. *Paradisi in Sole Paradisus Terrestris.* London: Methuen, 1904.

Platnauer, Maurice, trans. *Claudian Vol. 1.* Cambridge: Harvard University Press, 1922.

Plumwood, Val. *Environmental Culture: The Ecological Crisis of Reason.* London: Routledge, 2002.

———. *The Eye of the Crocodile.* Edited by Lorraine Shannon. Canberra: ANU E-Press, 2012.

———. *Feminism and the Mastery of Nature.* London and New York: Routledge, 1993.

———. "Nature in the Active Voice." *Australian Humanities Review* 46 (2009): 113–29.

Pritzke, Barry. *A Native American Encyclopedia: History, Culture, and Peoples.* Oxford: Oxford University Press, 2001.

Pughe, William Owen. "The *Mabinogion,* or *Juvenile Amusements,* Being Ancient Welsh Romances." *The Cambrian Register* (1795): 177–87.

Rackham, Harris, trans. *Pliny: Natural History Volume* 4. London: William Heinemann, 1960.

———, trans. *Pliny: Natural History Volume* 5. London: William Heinemann, 1938.

Reilly, Michael. "John White: The Making of a Nineteenth-Century Writer and Collector of Maori Tradition." *New Zealand Journal of History* 23 (1989):157–72.

———. "Review of *Traditional Stories from Southern New Zealand: He Kōrero nō Te Wai Pounamu,* translated and edited by Christine Tremewan." *New Zealand Journal of History* 38 (2004): 311–12.

Rieu, Emile Victor, and Dominic Christopher Henry Rieu, trans. *Homer: The Odyssey.* London: Penguin Books, 2003.

Roberts, Alexander, and James Donaldson, eds. *The Ante-Nicene Fathers: Translations of The Writings of Our Fathers down to A.D. 325 Volume VIIII.* Buffalo: The Christian Literature Company, 1886.

Roberts, Brynley. "Tales and Romances." In *A Guide to Welsh Literature, Volume One*, edited by A. O. H. Jarman and Gwilym Rees Hughes, 203–43. Cardiff: University of Wales Press, 1992.

Robinson, George, trans. *The Life of St. Boniface by Willballd*. Cambridge: Harvard University Press, 1916.

Rocher, Ludo. *The Puranas*. Wiesbaden: Otto Harrassowitz Verlag, 1996.

Rose, Deborah. *Country of the Heart: An Indigenous Australian Homeland*. Canberra: Aboriginal Studies Press, 2002.

———. *Dingo Makes Us Human: Life and Land in an Aboriginal Australian Culture*. Cambridge: Cambridge University Press, 1992.

———. "A Social and Ecological Analysis of Dreaming Trees." Unpublished manuscript, 1987.

Ross, William David, ed., *Nicomachean Ethics*. Oxford: Clarendon Press, 1908.

Rouse, William Henry Denham, trans. *Nonnus, Dionysiaca. Vol. 1.* Cambridge: Harvard University Press, 1940.

Roys, Ralph L., trans. *The Book of Chilam Balam of Chumayel*. Washington, DC: Carnegie Institution, 1933.

Sahlins, Marshall. "What Kinship Is (Part One)." *Journal of the Royal Anthropological Institute* 17 (2011): 2–19.

Schultes, Richard Evans, Albert Hofmann, and Christian Rätsch. *Plants of the Gods: Their Sacred, Healing, and Hallucinogenic Powers*. Rochester, VT: Healing Arts Press, 1992.

Seaton, Robert Cooper, trans. *Apollonius Rhodius: The Argonautica*. London: William Heinemann, 1912.

Segal, Robert. *Myth: A Very Short Introduction*. Oxford: Oxford University Press, 2004.

———. *Theorizing About Myth*. Boston: University of Massachusetts Press, 1999.

Seton, Kathryn, and John Bradley. "When You Have No Law You Are Nothing: Cane Toads, Social Consequences, and Management Issues." *The Asia Pacific Journal of Anthropology* 5 (2004): 205–25.

Shaw, George Bernard. "Back to Methuselah." http://www.gutenberg.org/ebooks/13084; accessed February 23, 2016.

Shorter, David Delgado. *We Will Dance Our Truth: Yaqui History in Yoeme Performances*. Lincoln: University of Nebraska Press, 2009.

Skene, William Forbes, trans. *The Four Ancient Books of Wales*. Edinburgh: Edmonston and Douglas, 1868.

Staal, Frits. *Discovering the Vedas: Origins, Mantras, Rituals, Insights*. London: Penguin, 2009.

Stanner, William Edward Hanley. *White Man Got No Dreaming: Essays 1938–1973*. Canberra: ANU Press, 1979.

Stasch, Rupert. *Society of Others: Kinship and Mourning in a West Papuan Place*. Berkeley: University of California Press, 2009.

Stirling, Matthew W. "Origin Myth of Acoma and Other Records." *Smithsonian Institute, Bureau of American Ethnology Bulletin* 135 (1942): 1–123.

Stone, Christopher D. "Should Trees Have Standing? Towards Legal Rights for Natural Objects." *Southern California Law Review* 45 (1972): 450–501.

Stoneman, Richard, trans. *The Greek Alexander Romance*. Harmondsworth: Penguin, 1991.

Storms, Godfrid. *Anglo Saxon Magic*. s'Gravenhage: Martinus Nijhoff, 1948.

Strathern, Marilyn. *Kinship, Law, and the Unexpected: Relatives Are Always a Surprise*. Cambridge: Cambridge University Press, 2005.

Suarez, Thomas. *Early Mapping of Southeast Asia: The Epic Story of Seafarers, Adventurers, and Cartographers Who First Mapped the Regions between China and India*. North Clarendon, VT: Tuttle, 2013.

Tacon, Paul S. C. "Identifying Sacred Landscapes in Australia: From Physical to Social." In *Contemporary Archaeology in Theory: The New Pragmatism*, edited by Robert W. Preucel and Stephen A. Mrozowski, 77–91. Chichester: Blackwell, 2010.

Tedlock, Dennis. *Two Thousand Years of Mayan Literature*. Berkeley: University of California Press, 2010.

Thorpe, Benjamin, trans. *The Poetic Edda*. Lapeer, MI: The Nothvegr Foundation Press, 2004.

Thulesius, Olav. *Nicholas Culpeper: English Physician and Astrologer*. New York: St. Martin's Press, 1992.

Tobienne Jr., Francis. *Mandeville's Travails: Merging Travel, Theory, and Commentary*. Lanham, MD: Rowman and Littlefield, 2016.

Tolstoy, Nikolai. *The Oldest British Prose Literature: The Compilation of the Four Branches of the Mabinogion*. Lampeter: Edwin Mellen, 2009.

Townshend, Richard Baxter, trans. *Tacitus: The Agricola and Germania*. London: Methuen, 1894.

Tunbridge, Dorothy. *Flinders Ranges Dreaming*. Canberra: Aboriginal Studies Press, AIAS, 1988.

Tylor, Edward. *Primitive Culture*. London: John Murray, 1920.

Van Buitenen, Johannes Adrianus Bernardus, trans. *The Mahābhārata Vol. 1 Book 1*. Chicago: University of Chicago Press, 1973.

Van Voorst, Robert E. *Anthology of World Scriptures*. Boston: Cengage Learning, 2016.

Vento, Urpo. "The Role of the Kalevala in Finnish Culture and Politics." *Nordic Journal of African Studies* 1 (1992): 82–93.

Victoria University of Wellington Library. "Ancient History of the Māori." http://nzetc.victoria.ac.nz/tm/scholarly/tei-corpus-WhiAnci.html; accessed February 23, 2017.

Vijñanananda, Swami, trans. *The S'rîmad Devî Bhâgawatam*. 1921–22. sacredtexts.com; accessed January 31, 2016.

Voloshinov, Valentin. *Marxism and the Philosophy of Language*. Cambridge: Harvard University Press, 1986.

Von Simson, George. "The Mythic Background of *The Mahābhārata*." *IT* 12 (1984): 191–223.

Wallis Budge, Ernest Alfred, trans. *The Book of the Dead: The Chapters of Coming Forth by Day*. London: Kegan Paul, Trench, Truebner, 1898.

———, trans. *The Egyptian Texts*. London: Kegan Paul, Trench and Trübner, 1912.

Waltke, Bruce K. "The Creation Account in Genesis 1:1–3, Part IV: The Theology of Genesis 1." *Bibliotheca Sacra* 132 (1975): 327–42.

Warner, Arthur George, and Edmond Warner. *The Shahnama of Firdausi*. London: Kegan, Paul, Trench, Truebner, 1905.

———, trans., *The Shahnama of Firdausi Vol 6*. London: Kegan Paul, Trench, Truebner, 1912.

Watson, James et al. "Catastrophic Declines in Wilderness Areas Undermine Global Environment Targets." *Current Biology* 26 (2016): 2929–34.

Werner, Edward Theodore Chalmers. *Myths and Legends of China*. London: George G. Harrap, 1922.

West, Edward William, trans. *Pahlavi Texts Part* 1. Oxford: Clarendon Press, 1880.

West, Martin Litchfield. "Odyssey and Argonautica." *The Classical Quarterly* 55 (2005): 39–64.

White, John. *The Ancient History of the Maori, His Mythology and Traditions Volume 1*. Wellington: George Didsbury, Government Printer, 1887.

White Jr., Lynn. "The Historical Roots of Our Ecologic Crisis." *Science* 10 (1967): 1203–07.

Wilson, Horace Hayman, trans. *The Vishnu Purana*. London: John Murray, 1840.

Witzel, Michael. "The Development of the Vedic Canon and its Schools: The Social and Political Milieu." *Harvard Oriental Series, Opera Minora* 2 (1997): 257–348.

Wolohojian, Albert Mugrdich. trans. *The Romance of Alexander the Great by Pseudo-Callisthenes*. New York: Columbia University Press, 1969.

Woodruff, John. "Ma(r)king Popol Vuh, *Romance Notes* 51 (2011): 97–106.

Xiaodong, Wu. "The Rhinoceros Totem and Pangu Myth: An Exploration of the Archetype of Pangu." *Oral Tradition* 16 (2001): 364–80.

Yusoff, Kathryn. "Aesthetics of Loss: Biodiversity, Banal Violence, and Biotic Subjects." *Transactions of the Institute of British Geographers* 37 (2012): 578–92.

poplar, 73, 89, 147, 203
Popol Vuh, xxix, 7–8, 25, 70, 78, 146–147,
    236–237
Pouteria lucma. See lucma
primeval ox, 5, 18, 233–234
Prose Edda, 4, 231–232
Pseudodoxia Epidemica. *See* Browne, Sir
    Thomas
Puranas, xxxix, 6, 39–40, 56, 192, 198, 212,
    235–236
Puruṣa, 4, 16

Rangi-nui, 9, 30
Rātā, 194–195, 213–215, 240
red lily, 76, 107. *See also* sacred lotus
reincarnation, 69, 75, 103, 190, 247, 252
Rem, 7
rice. *See Oryza sativa, Oryza perennis*
Rig Veda, 4, 16, 232–233
ritual, 228–229
rowan, 22
rudraksha, 40, 56

sacred lotus, 40, 58, 236
sacredness, 35–38, 145, 159–161, 198
saka-ki, 42, 63
salmali, 150, 181–185
sap, 81, 89, 117, 128, 145–148, 172
Saussurea costus. See kushtha
Scythian Lamb, 112, 122
sentience, 38, 40–43, 74, 113, 140–190
Shahnama, 149, 175–177, 251–252
shared substance, 3–8, 12, 40, 68–107, 110–
    112, 197
Shiva, 39–40, 56
soma, 43, 65–67, 115, 137, 191, 198. *See also*
    haoma
speech, 145, 147–150, 153, 158–161, 165–
    169, 175–185, 197, 229
spirit, 13, 26, 66, 70, 116, 151, 219, 233
Star of Bethlehem. See *Ornithogalum umbellatum*
suffering, 38–39, 50, 72, 145–152, 163–168, 189
sweat, 4, 6, 13, 20, 40, 56, 164
symbolism, xxv–xxvi, 35–40, 53, 73–74, 115–
    117, 144, 228

talking to plants, 117–119, 168–169, 171–174
Tāne, xxx, 9–11, 30, 194–195, 215, 240
tears, 7, 73–74, 81, 86, 89, 94, 144, 147, 152,
    171, 176
Theophrastus, 111
thunder tree, 197, 225
Tiger Shark, 9, 27, 239
Tillandsia usneoides, 6
Tlingit, 77
tobacco plant, 42, 206
Tonga, 74, 100–101
tōtara, 11, 30, 194
tree of the knowledge of good and evil, 36, 44
trees of sun and moon, 149, 179–180
Tsichtinako, 8, 26, 191, 194, 205–206, 238
tulsi, 39–40, 56

Upaniṣads, xxix, 5, 36, 75, 103, 190, 247–248

Väinämöinen, 21–22, 38, 51, 148, 171–174,
    192, 209–211, 234
violence, 41–42, 95–101, 145–148, 159–161,
    168–169, 189–225
Virgil, xxix, 71, 146–147, 165–167, 244
Vishnu, 39, 40, 56
volition, 143, 155–157
Völuspá, 5, 36, 48, 70, 231–232, 234

wak-wak, 113, 124, 252
Warramurrungundji, 9, 27, 239
whakapapa, 9, 10–11, 240
world trees, 36–39, 48–55, 115
worship, 34–36, 38–40, 83, 91

Yanyuwa, 9, 12, 27, 238–239
Yaqui, 149–150
yas, 116
yaxche, 37–38, 53–55
Yggdrasil, 36–39, 48–50, 232
Yilig-moi-indih, 76
Ymir, 4–5, 9, 13

Zeus, 41, 61, 73, 89, 187
Ziziphus spina-christi, 143, 153
Zuni, 8, 243